THOMAS O'CONOR SLOANE

LIQUID AIR
AND THE
LIQUEFACTION OF GASES

Elibron Classics
www.elibron.com

Elibron Classics series.

ISBN 1-4021-6017-8 (paperback)
ISBN 1-4021-0427-8 (hardcover)

This Elibron Classics Replica Edition is an unabridged facsimile of the edition published in 1920 by the Norman W. Henley Publishing Co., New York.

LIQUID AIR
AND THE
LIQUEFACTION OF GASES

A PRACTICAL WORK GIVING

The entire history of the liquefaction of gases from the earliest times of achievement to the present day—the biography of the great investigators—the manipulation of liquid air and liquefied gases —experiments with the same—the modern uses of liquefaction processes and of their products—the utilization of the nitrogen and oxygen of the air, the rare gases, helium, argon, and neon, and their utilization, and the latest developments in this field of industrial science, are also treated.

BY

T. O'CONOR SLOANE, Ph.D.

Author of "Arithmetic of Electricity," "Electricity Simplified," Etc.

FULLY ILLUSTRATED

THIRD EDITION—REVISED AND MUCH ENLARGED

NEW YORK

THE NORMAN W. HENLEY PUBLISHING CO.

2 West 45th Street

1920

PRINTED IN U. S. A.

Composition, Electrotyping and Press Work
By The Publishers Printing Company, New York

PREFACE TO THIRD EDITION.

Over twenty years have elapsed since air was liquefied on the large or manufacturing scale. It was handled in ordinary containers, dippers, tumblers or barrels just like water, and with far less precaution than is exacted by the acids of commerce. As it inevitably evaporated and if confined would develop an enormous pressure, the vessels employed to contain it were necessarily open. The story of liquid air and of the liquefaction of other gases reads like a fairy tale. But the world has become so satiated with miracles of science that it has almost forgotten how to wonder.

In Gulliver's veracious account of his travels we read of the work done in the famous Academy of Logado. In one department fifty men were at work under the superintendence of the universal artist, as one of the illustrious investigators was called. These men were engaged in various occupations. "Some were condensing air into a dry tangible substance by extracting the niter, and letting the aqueous or fluid particles percolate." So says the great Dean, selecting the solidification of air as one of the impossibilities worthy of embodiment in his sarcastic romance.

At first when the manufacture of liquid air on the large scale was developed the question arose of what to do with it, for it seemed that so wonderful a thing must have its uses. It was found that the nitrogen escaped by evaporation much more rapidly than did the oxygen, and in this selective evaporation was seen a basis for the manufacture of oxygen, letting the supposedly useless nitrogen go to waste. This saved at the best one-fifth of the liquid. Now it is the nitrogen which is wanted, and the oxygen is saved as a by-product, and about four-fifths of the air are utilized in the production of what has become the main product.

The present work aims to tell the history of the liquefaction of gases, wherein the physicist has exceeded the fictitious achievement told of in Gulliver's Travels. The subject, whose history extends over more than a century, is full of interest from the biographical and historical as well as scientific standpoint, and it is hoped that the presentation of it from such viewpoints will be acceptable to the reader.

For assistance in the compilation of this work the author's thanks are due to more than one investigator. His requests for information met with quick response from such men as Raoul Pictet, L. P. Gailletet, Henri Dufour, Charles E. Tripler and James Dewar.

It is especially to these pioneers in the difficult paths of investigation that his and the reader's grateful acknowledgments must be given.

OCTOBER, 1919.

TABLE OF CONTENTS.

CHAPTER I.—PHYSICS. Pages 9–36

What is liquid air?—The three states of matter: Solid, liquid and gaseous—Relations of pressure and heat to state assumed by matter—The critical state and its phenomena—Natterer's tube—Physical units—Space, mass and time—Force and energy—Conservation of force an abandoned doctrine—Conservation of energy—Work a synonym for development of energy—Waste of energy and entropy—Possibility of utilizing the lower forms of energy of the universe.

CHAPTER II.—HEAT. Pages 37–58

Heat and its measurement—Thermometers—The zero point—The Celsius or Centigrade thermometer scale—Fahrenheit's thermometer scale—The absolute zero—Its basis—Coefficient of expansion of gases—Determination of temperatures in the liquefaction of gases—Different liquids used in filling thermometers—The air thermometer—The hydrogen thermometer—Details of its construction—Electrolytic hydrogen—The hydrogen or air thermometer formula—The thermo-electric thermometer—Onnes' instrument and details of its construction—Its calibration—The electric resistance thermometer—Calorimetric determination of temperatures.

CHAPTER III.—HEAT AND GASES. Pages 59–84

The perfect gas—The ultra-perfect gas—Energy expended in heating a gas—Specific heat at constant pressure and at constant volume—Atomic heats and variations of same from equality with each other—Adiabatic and isothermic expansion of gases—Carnot's cycle—The perfect heat engine—Available and unavailable energy—Unavailable energy rendered available by liquid air—Latent heat of melting, of vaporization, of expansion—Boiling a cooling process—Expansion a cooling process—The spheroidal state—The Crookes layer—Experiments and illustrations—Utilization of the spheroidal state in low temperature work and in liquid air investigations.

CHAPTER IV.—PHYSICS AND CHEMISTRY OF AIR. Pages 85–91

The atmosphere as an ocean—What air is—Its constituents—Relations of air to living beings—The chemist's and physicist's view of air—Its constancy of composition—Carbon dioxide—Oxygen—Nitrogen, argon and other constituents.

TABLE OF CONTENTS.

CHAPTER V.—THE ROYAL INSTITUTION OF ENGLAND. Pages 92–99

The Royal Institution—Its origin and objects—Count Rumford—Sir Humphry Davy—The Pneumatic Institute—Davy's experiments in inhaling poisonous gases—His engagement as director of the Royal Institution—His views on the utility of liquefying gases.

CHAPTER VI.—MICHAEL FARADAY. Pages 101–115

Michael Faraday—His early life—Early devotion to science—His introduction to Humphry Davy—Attendance at scientific lectures—Engagement at the Royal Institution—Injuries from explosion in the laboratory—European tour with Davy—Rivalry of scientific men—Davy and Faraday as rivals—The liquefaction of chlorine—Davy's share in the experiment—Davy's opposition to Faraday's election as fellow of the Royal Society—Dr. Paris and the liquefaction of chlorine—Faraday's descriptions of his liquefactions—Explosions—Northmore's priority published by Faraday—Notes on Faraday's liquefaction of gases—His exhibit on of Thilorier's apparatus—His later work in liquefying gases—Discovery of the magnetism of oxygen gas—His death—Bent tubes as used by Faraday—Experiments with use of bent tubes—The Davy-Faraday Laboratory.

CHAPTER VII.—EARLY EXPERIMENTERS AND THEIR METHODS. Pages 116–151

Perkins' claim to have liquefied air—Its absurdity—Northmore's liquefaction of chlorine—Rumford's experiments as commented on by Faraday—Babbage's experiment in a drill hole in limestone rock—Monge and Clouet's alleged liquefaction of sulphurous oxide—Faraday's liquefaction of chlorine—Stromeyer's liquefaction of arseniureted hydrogen—Faraday's bent tubes for liquefaction of gases—Manometer for use with them—Experiment in a straight sealed tube on the liquefaction of chlorine—Davy's suggested method—Cagniard de la Tour—His bent tube experiments—D. Colladon—His apparatus as still preserved—Thilorier—His discovery of solid carbon dioxide—A fatal explosion—The improved Thilorier apparatus—Johann Natterer's apparatus—His experiments—Loir and Drion's solidification of carbon dioxide—Thomas Andrews, of Belfast.

CHAPTER VIII.—RAOUL PICTET. Pages 153–171

The life of Raoul Pictet—His education—His ice machines—Disputed priority—Honors awarded—His apparatus for liquefying gases—Description of its operation—Temperatures of the cycles of operation—His dispatch of December 22, 1877, to the French Academy—Regnault's statement—Hydrogen—His dispatch of January 11, 1878, to the French Academy - Olszewski's comments on the hydrogen experiment—Pictet's arrangement of pumps—His desire to produce liquid oxygen in quantity—Comments on his work—The *liquide Pictet*.

CHAPTER IX.—LOUIS-PAUL CAILLETET. Pages 173–202

The life of L.-P. Cailletet—His education—Honors received—His modification of Colladon's apparatus—Accidental liquefaction of acetylene by release—Description of his apparatus—How the apparatus was filled—The full apparatus with hydraulic press—Liquefactions of nitrogen oxide—Of carbon monoxide and oxygen mixed—Liquefactions of the same separately—His letter of December 2, 1877, to the French Academy—Liquefaction of nitrogen—Of hydrogen—Rival claims of Cailletet and Pictet—Mercury stopper method—Manometers—Original methods of testing—Eiffel tower manometer—Carbon dioxide experiments—Mercury pump—High pressure gas reservoir—Ethylene as a refrigerant—Closed cycle method—Accelerated evaporation—Electric conductivity at low temperatures—Comparison of thermometric methods—La Tour's experiment repeated.

CHAPTER X.—SIGMUND VON WROBLEWSKI AND KARL OLSZEWSKI. Pages 203–229

Wroblewski's life—Banishment from his native country—Early scientific work—His association with Olszewski—Study of Cailletet's methods—Their apparatus—Defective position of the hydrogen thermometer—Liquefactions of oxygen, carbon monoxide and nitrogen—Ethylene data—Solidification of carbon disulphide and alcohol—Determination of the critical pressure and temperature of oxygen—Liquefaction of hydrogen—Use of a thermo-electric thermometer—Electric resistance of metals at low temperatures—Two liquids from air—Olszewski's individual work—Apparatus for producing liquid oxygen in quantity—Comparison of platinum resistance and of hydrogen thermometers—Determination of hydrogen constants.

CHAPTER XI.—JAMES DEWAR. Pages 231–285

Dewar's life and education—His associates—Controversies with Cailletet as to priority—Early liquefaction apparatus—Solid nitrous oxide as a refrigerant—Royal Institution apparatus—Cooling cycles employed—Laboratory apparatus—Vacuum vessels—Air as a heat conveyer—Experiments with incandescent lamps—Reflection of ether waves from vacuum vessel—Keeping power of vacuum vessels—The Dewar vacuum—Its extraordinary perfection—Analogy with population of earth—Experiment in slow diffusion of mercury vapor—Incidental production of vacuum vessels—Elasticity and strength of metals at low temperatures—Apparatus used—Elongation of metals when stressed at low temperatures—Determination of specific and latent heats of liquefied gases—Gas-jet experiments—Low temperatures thus obtained—Freezing air—Large jet apparatus—Analysis by liquefaction—Liquefaction of fluorine—Liquefaction of hydrogen and helium—Experiments to show the intense cold of liquid hydrogen.

CHAPTER XII.—CHARLES E. TRIPLER. Pages 287–296

The life of Charles E. Tripler—His early experiments with gas motors—Mechanical difficulties encountered—His electrical experiments—Chemistry—His work in fine art—Exhibition of his paintings—Return to the investigation of compressed gases—Liquefaction of air—He endeavors to utilize the low grade heat of the universe—Simplicity of his apparatus—The plant—The compressor—General plan of operations—Capacity of his plant—How he transports liquid air—His lectures—Raoul Pictet in Charles E. Tripler's laboratory.

TABLE OF CONTENTS.

CHAPTER XIII.—THE JOULE-THOMSON EFFECT.

Pages 297-306

First attempts at liquefying gas—Joule and Thomson and their discovery—Coal a cheap chemical—Substitution of mechanical for chemical energy—Sir William Siemens' regeneration of cold—Self-intensive refrigeration—Negative Joule-Thomson effect—Mathematics of the theory—Conditions of pressure for economical application.

CHAPTER XIV.—THE LINDE APPARATUS.

Pages 307-319

Linde's apparatus—The simplest form of apparatus—Its operation—Its storing of air at atmospheric pressure—Avoidance of atomization and waste—Subdivision of pressure-drop—Laboratory apparatus—A feature of inefficiency in it—Its power of liquefaction—Continuous oxygen-producing apparatus—Date of Linde's first successful use of his apparatus.

CHAPTER XV.—THE HAMPSON APPARATUS.

Pages 320-324

Hampson's apparatus—Its general features of construction—The jet and regulating device—Thermal and mechanical advantages—Data of its operation—Use of cylinders of compressed gas instead of pumps—Application of preliminary cooling to the air or gas to be liquefied.

CHAPTER XVI.—EXPERIMENTS WITH LIQUID AIR.

Pages 325-337

Experiments with liquid air—Formation of frost on bulbs—Filtering liquid air—Dewar's bulbs—Liquid air in water—Tin made brittle as glass—India rubber made brittle—Descending cloud of vapor—A tumbler made of frozen whisky—Alcohol icicle—Mercury frozen—Frozen mercury hammer—Liquid air as ammunition—Liquid air as basis of an explosive—Burning electric light carbon in liquid air—Burning steel pen in liquid air—Carbon dioxide solidified—Atmospheric air liquefied—Magnetism of oxygen.

CHAPTER XVII.—SOME OF THE APPLICATIONS OF LOW TEMPERATURES. Pages 338-356

Frigotherapy—The frigorific well—Pictet's experiment—Effects of the first trial of the system—Medical uses of liquid air—Critical point as test of purity of chemicals—Purification of chemicals by low temperature crystallization—Low temperature distillation—Regulation of chemical reactions by cold—Liquid air explosives—The principle of their action—Liquid air in electric power transmission—Liquid air as a reservoir of energy.

TABLE OF CONTENTS.

CHAPTER XVIII.—CLAUDE AND LINDE. Pages 358-375

Nitrogen of the Air and Its Utilization—Georges Claude—New Importance of Oxygen—Perfect Gases—Dewar's Cycles—Linde's Principle—Low Temperature Regeneration—Claude's Investigations and Their Developments—Gasoline, a Low Temperature Lubricant—Interchangers—Claude's Apparatus—Relations of Oxygen and Nitrogen in Liquefaction and Evaporation—Dew Point Production of Liquid Gases—Retrograde Condensation or Backward Return — Rectification—Claude's Oxygen Apparatus—Linde's Nitrogen Apparatus—Claude's Nitrogen Apparatus.

CHAPTER XIX.—UTILIZATION OF ATMOSPHERIC GASES.
Pages 376-378

Cyanamid Process for Utilization of Atmospheric Nitrogen—Calcium Carbide as Used in the Cyanamid Process—Direct Use of Cyanamid as a Fertilizer—Cyanid Process for the Utilization of Atmospheric Nitrogen—Heber's Process for the Same—Literature of the Nitrogen Industries.

CHAPTER XX.—HELIUM, ARGON AND VARIOUS NOTES.
Pages 379-385

Charcoal Absorption of Gases at Low Temperatures—Charcoal of Highest Absorbing Power for Gases—Kammerlingh-Onnes' Liquefaction of Helium—The Apparatus Used—Use. of Helium for Balloons—Use of Argon for Incandescent Lamps—Liquid Hydrogen Vacuum—Variation of Liquefaction Temperatures of Air—Chemical Affinities at Low Temperatures—Liquid Ozone—Blasting Cartridges from Liquid Air.

LIQUID AIR

AND THE

LIQUEFACTION OF GASES

CHAPTER I.

PHYSICS.

What is liquid air?—The three states of matter: Solid, liquid, and gaseous—Relations of pressure and heat to state assumed by matter—The critical state and its phenomena—Natterer's tube—Physical units—Space, mass, and time—Force and energy—Conservation of force an abandoned doctrine—Conservation of energy—Work a synonym for development of energy—Waste of energy and entropy—Possibility of utilizing the lower forms of energy of the universe.

A question has often been asked latterly; it is, "What is liquid air?" The subject has been so much discussed, and so much has been made of it, that it is hard to believe that there is not some occult mystery attending it. Liquid air is simply air which is so cold that it assumes the liquid state.

The fact that the question has been so often asked suggests the need for a thorough answer; for back of it there lies a great region of physics and chemistry; a summary exploration of which in the light of the knowledge of to-day cannot but be interesting. In it are concerned the great doctrine of the conservation of energy, the laws of heat, the three states of matter, and the chemistry of air, and it is not expecting too much of the reader of to-day to hope that the theory of the subject presented within the compass of an hour's reading will interest him.

The account of the liquefaction of gases includes a period of about one hundred years, and with it is bound up the history of the Royal Institution of London. In its laboratory Faraday worked with bent tubes, liquefying gases and blowing the tubes to pieces and nearly blinding himself in his efforts. This was half a century ago and more. And now within its walls, with elaborate machinery based upon Pictet's circuits of 1877, James Dewar, the successor of Faraday, liquefies hydrogen and helium and ends the century's work.

In Switzerland and France, toward the end of 1877, the beginning of the end appeared when oxygen was liquefied. Pictet and Cailletet were the rivals, separated only a few days in their liquefaction of this gas, discovered by Priestly and Lavoisier almost exactly one hundred years before the date of its reduction to the liquid state.

America was not idle. Tripler working away privately, with no institution or association to back him, has surpassed the dreams of the most enthusiastic visionaries and has made liquid air by the barrel, and has sent it all over a wide range of country in tin cans.

The long record should not be read until the answer to the query cited above has been given; the reader should know accurately what liquid air is, what constitutes a gas, what the relations of heat and pressure to state of matter are, and how heat is treated by the modern scientist.

Matter is generally stated to exist in three forms or states—the solid, liquid and gaseous. An attempt has been made to assert the existence of a fourth state—the ultra-gaseous or radiant state. There is a certain objection, however, to this. The first three states are broadly differentiated. As a rule, there is little question of the form or state being solid, liquid or gaseous, but the ultra-gaseous state is only recognizable by rather refined tests and may perhaps be better considered as the extreme carrying out of the gaseous condition.

Water is the most convenient substance to cite to illustrate the three states. In ice we have solid water. The masses are of fixed contour, and, even if ice is subject to a species of flow, the masses of ice definitely hold their shape. The molecules of solid water are in constant vibration back and forth over the same path, under any conditions of temperature we are familiar with. At the absolute zero this motion would cease. The paths are inconceivably short. We cannot and probably never will acquire any direct knowledge or sight of these vibrations. All we know is that ice at mundane temperatures is hot. It will be seen that, dropped into liquid air, it makes it boil as if the ice were a red hot poker thrust into it. By the kinetic theory of heat all hot bodies are held to have their molecules in constant vibration. Molecular attraction holds the particles of the ice firmly together in spite of this vibration.

If we apply heat, we diminish this attraction, we increase the repulsive forces, and the ice reaches a temperature where the two opposing forces about balance each other, the attractive ones slightly preponderating. Now there is no longer a powerful set of forces in operation binding the molecules together. They begin to slide about on each other, their vibrations continue with energy, but the paths vary. A molecule bounces back and forth like a billiard ball, recoiling to right or left from its neighbor, so that sooner or later it travels through the entire mass and never ceases its travels. As the molecules slide about without true friction the ice loses all tendency to preserve its shape and falls to pieces, literally speaking. In other words, the ice melts, and we have water—a representative of the liquid state of matter.

Let us apply more heat. Our water is already shapeless. We have to keep it in a containing vessel. Even a drop of water hanging from the window shutter on a rainy day is held in a little sack of water-film. Later on we shall see what an important bearing the liquid film has in the manipulation of liquid air. So we put our water in a kettle and heat it. Soon a white cloud issues from the spout, and we may say that we see the steam. If we make such an assertion, it is an erroneous one, as the white cloud is really composed of little balls of liquid water, each held in its own little sack of water-film. As the kettle boils harder, we find that the white cloud does not begin its existence until it is a few inches from the mouth of the spout, and a space apparently void of all matter intervenes between spout and white cloud. This space is filled with the substance we are in search of; it is occupied by a column of gaseous water or steam rushing out of the spout and as invisible as air itself.

By applying heat to our water, we have made the molecules vibrate through paths many times longer than the old paths; a cubic inch of water gives us approximately a cubic foot of steam. The molecules travel about through the mass with greater rapidity than ever. The mass loses all pretensions to shape or cohesion. A vessel will not hold it unless it is closed everywhere. The third state of matter is formed—the water exists as a gas.

By refinement of observation and experiment most interesting and captivating views are formed concerning these states of matter. Their individual properties are not so sharply cut off and defined as might be supposed. A body is said to be solid when it is practically unchanging in the shape imparted to it. But many solids flow under pressure. The suffering "continuous deformation under the action of a continuous force" is not a certain criterion of a liquid, but it is good enough to define it or identify it by.

A barrel of asphalt opened and thrown on its side in the street seems to be filled with a black solid, yet by the end of the day it will have flowed and changed shape. A stick of sealing wax supported at its ends slowly and continuously bends. Some authorities consider these as examples of liquids.

A soft jelly pressed by a spoon yields considerably, but, when the pressure ceases, springs back into its original shape. Jelly, therefore, is treated as a solid.

All this seems to cast confusion on the subject.

But nothing very critical hinges on the sharp separation of solid, liquid and gas. It would perhaps be better to assume a continuity of state between solids and liquids, and to consider asphalt, sealing wax and the like as being on the border line. If sealing wax is to be considered a liquid, then lead and most other metals could be considered such; for metals, as a rule, are more or less malleable and ductile, and the qualities of malleability and ductility depend upon the flow of the material composing them.

We are confronted with the old property of nature expressed in the adage, *Natura non facit saltum*, Nature does not jump. The air we breathe is in the gaseous condition. The water we drink is in the liquid condition. The glass which holds the water is in the solid condition. Yet we can indicate many cases where an intermediate state exists and where a substance cannot well be termed one thing or the other. Even air is not a perfect gas, and hydrogen is an ultra-perfect gas.

For want of correct understanding of such things as these, confusion in ideas results and an obscurity bordering upon complication is introduced into our conception of the laws and system of nature. Thus moist air is generally considered heavier than dry air, presumably because a wet cloth is heavier than a dry one. Popularly, people would say that the air is damp and heavy. Now air is wet because of the mixture with it of another gas, gaseous water or literally steam. Water from rain, from the ground and from the immense evaporating surface of the leaves of the vegetable world assumes the gaseous form and mixes with the air. The specific gravity of water in the gaseous condition is less than that of air. It is about two-thirds as heavy only. Wet air, therefore, is lighter than dry air. A balloon would rise better on a dry day than on a wet day, not only because there would be no moisture with which to dampen the cordage and cloth, and thereby increase the weight, but because the dry air is a better floating medium than wet air, because it is heavier.

Wet air is not air soaked like a sponge with water. It is simply a mixture of dry air with gaseous water. The truth is here far simpler than fiction.

The sequence followed by a substance in passing from state to state is not always the same, as a solid on heating is often vaporized or gasified directly without passing into the liquid state at all. This occurs in slow vaporization very often. Thus ice in the open air below the freezing temperature wastes away by volatilization and is gasified slowly, without liquefying, and contributes water vapor to the air, although far below the solidifying temperature. Iodine volatilizes in the same way, and those who have used camphor or naphthaline for preserving clothes from moths have observed the same mysterious diminishing of the lumps of preservative used. In druggists' windows the shrinkage of camphor there exposed is sometimes quite striking. Now it is less often exposed than formerly, as naphthaline has largely supplanted it in the trade.

Carbon dioxide, the gas which escapes from soda water and other effervescent beverages, when subjected to cold and pressure, liquefies. When the pressure is released and it is allowed to escape into the open air, it solidifies and produces a true carbon dioxide snow. This snow exhibits surprising permanency, disappearing quite slowly in the open air. In disappearing it evaporates and produces gas directly without passing through the intermediate liquid state.

Such direct transition from a solid into a gaseous state is termed often sublimation; an expression, perhaps, too limiting, covers the extreme case where a solid on application of heat sublimes vigorously before melting. It is to the effect that the substance boils at a lower temperature than that at which it liquefies—that the temperature of boiling is lower than that of liquefaction. The idea of a solid boiling seems rather odd.

It is not only the change of temperature which brings about change of state. Change of pressure affects it greatly. The greater the pressure, the higher is the temperature at which a liquid becomes a gas. A gas just hot enough to hold that form may, under some conditions, be converted into a liquid by applying pressure, without any change in temperature being required to effect the change of state. This, too, is very natural. For a liquid, under ordinary conditions, being of smaller volume than the same molecules gasified, is naturally brought to the liquid condition by mechanical reduction of volume as well as by thermal reduction.

Pressure will not always do it, and by combining the effects of great heat and great pressure, conditions foreign to the ordinary status of matter are brought into existence which complicate the problem. Heat is the great and all-controlling agent. Heat is what establishes the critical state, and pressure is quite a secondary matter. For every gas there is a critical temperature and a critical pressure, but the latter is quite a subsidiary thing, and is not critical in the full sense that the temperature is.

Pressure tends to liquefy a solid, if the latter grows smaller on liquefaction. So that it is quite conceivable that a point might be reached where pressure would help to convert a liquid into a gas. As such a phenomenon, uncomplicated by other factors (page 24), has never been observed, it is better to set it aside and consider pressure as invariably on the side of cold in liquefying gases.

A gas must be pictured to the imagination as a very active thing. In a room full of air the molecules are moving about rapidly, colliding with each other, and bounding about like billiard balls. We know that, if we turn on the gas without lighting it, in a very few minutes the odor of gas will be perceived in all parts of the room. This can only be so because in those few minutes the gas has penetrated every corner. Its molecules have traveled about until some of them are everywhere present, and the activity of their operations may be judged by the amount of gas and the size of the room. An ordinary burner delivers one cubic foot of gas in about ten minutes, and in that time a room of over a thousand times that volume would be pervaded with it. Hence it will be seen how active the molecules of a gas are.

If there were no wind, if the air were absolutely motionless, its molecules would be as active as ever in their own spheres. The air which on one day would be in America would be scattered the next day far and wide, and its molecules would find their way sooner or later all over the world.

The same is true in a lesser degree of liquids. The water of a tideless, currentless lake is in molecular motion. The water which beats against the coast of America is in constant process of change, and its molecules are changing and moving about all the time. Sooner or later some of them will be in the waves which break upon the Irish cliffs and English beaches, nearly three thousand miles away. They would travel thus were there no oceanic currents and no waves.

This molecular travel is termed diffusion.

We have seen that the motions of the molecules are increased in vigor by heat, that, if heat is withdrawn, they decrease in intensity. The obvious question arises, What would happen if there were no heat? The molecular motions would cease, and molecular death would ensue.

The passage of a substance from the solid to the liquid state or from the liquid to the gaseous state involves generally a change in dimension or size, and in the case of many substances the liquid state is the one of smallest size. This is the case with water. In round numbers, a pint of water gives nearly a pint and two ounces of ice, if it freezes, and if converted into steam, gives nearly two hundred gallons. We are most concerned with the liquid and gaseous states, and under ordinary circumstances there is a very great reduction of volume incident to the passage of a substance from the gaseous to the liquid state.

It follows that, to produce liquefaction of a gas,

the first thing we should naturally try to do would be to reduce it in volume, and the simplest way to do this would be by pressure. Early experimenters adopted this plan. Natterer attained pressures of many thousand pounds to the square inch, yet gases compressed to a small fraction of their volume staid gases and refused to yield.

At last Andrews, of Belfast, made his classic discoveries, and the existence of a critical state was established. This state is very easy to understand. It depends on the fact that for every gas there is a temperature called its critical temperature, and a corresponding pressure called the critical pressure. When hotter than this temperature, no compression, however great, will liquefy it. Below this temperature, a compression easy of attainment is enough to effect the change to the liquid state.

The critical pressure is a term which is often misunderstood. It may be said that the pressure is never critical in the full sense in which temperature becomes critical. There is no pressure which can be defined as so low that liquefaction would be impossible in it. There is a theoretical point of cold never yet attained, which is termed the absolute zero. At this point heat ceases, the molecules no longer vibrate, and absolute cold exists. If a body were reduced to the absolute zero, where the motions of the molecules cease, pressure would be without effect upon it, as its only power is to shorten the paths of vibration of the molecules. The term critical pressure is used to describe the pressure required to liquefy a gas when it is at the critical temperature.

When a gas is at the critical temperature and at

the critical pressure also, the least increase of pressure or decrease of temperature will convert it into a liquid. When in this condition, ready to be a gas or a liquid, it is said to be in the critical state.

It will be seen how very well the term critical state applies when a substance is at the critical pressure and temperature, the least change will so profoundly modify its state.

A law relating to the critical state is known as La Tour's law, and expresses very succinctly the phenomenon of the critical temperature. It is the following:

There is for every vaporizable liquid a certain temperature and pressure at which it may be converted into the aeriform state in the same space occupied by the liquid.

It will be evident how strikingly this puts the fact that, above a certain temperature, a gas can be squeezed down to the volume of its mass as a liquid without liquefying. If a gas rigorously followed Mariotte's law and changed in volume in inverse proportion to the pressure exerted upon it, and if pressure sufficient to reduce it to the absolute volume, as it may be termed, or the volume it should have at the absolute zero, were exerted upon it, it is hard to say what would become of it.

The condition of a substance in the neighborhood of the critical state is sometimes termed the intermediate state. The expressions are almost synonymous—the first is the more abstract, the latter the more concrete expression.

The reduction from the gaseous to the liquid state is usually a reduction of volume. A cubic foot of steam gives about a cubic inch of water; eight hundred cubic inches of ordinary air give about a cubic inch of liquid air. But owing to the phenomenon of the critical temperature, or, what is the same thing, to La Tour's law, this is not always true. The existence of a gas of no greater volume than the liquid it could be converted into is a sort of scientific riddle. It has its counterpart in the inexplicably great power of expansion by heat possessed by some liquefied gases without departure from the liquid state.

The passage of a substance from the liquid to the gaseous state is marked by a change of appearance A liquid has always a defined limit. It lies in the containing vessel and its upper surface forms a visible boundary. If the vessel is of large diameter, the surface is level and flat, except along the edges, where it curves up or down a little. If the diameter is small, it curves throughout its whole extent, forming a little cup or a little hill, as the case may be.

The upper curved surface of a liquid is termed the *meniscus*. Mercury in a glass tube forms a convex meniscus; water, a concave one. For different liquids in contact with solids the meniscus varies, a characteristic one obtaining for each condition.

A very interesting suggestion is due to Jamin. It is that when oxygen and carbon dioxide are compressed together, a point may be reached when the carbon dioxide will liquefy but will be lighter than the compressed gas, so that we should have the curious phenomenon of a fluid floating upon a gas. Prof. Ramsay seems to think that he has observed this phenomenon. The meniscus in this case lies at the bottom of the liquid and above the gas.

For years the disappearance of the meniscus was regarded as marking the change or transition from the liquid to the gaseous state. This view seemed satisfactory. But science is not restful. Doubts began to be cast upon the coincidence of this disappearance with the true transition.

Thus in 1892 Zambiasi attacked the problem by experimenting with ether in a sealed tube and reproduced the intermediate and critical state phenomena therewith. Cagniard de la Tour's and Cailletet's observations were studied with the more manageable ether. Zambiasi came to the conclusion that the appearance and disappearance of the meniscus, while occurring at a constant temperature for a given tube, occurs at different temperatures in different tubes, the temperature being determined by the relative proportion of liquid to gas in the tubes.

In 1893 there were published a number of papers by Ramsay, Galitzine and others on the subject of the critical state and the uncertainty of the optical method, by simple inspection, of determining the transition from liquid to gaseous state. Quite an acrimonious discussion is contained in successive communications between the opposition scientists. The subject is left rather unsettled; the disappearance of the meniscus with some has lost its old time definite status, and the case is left pretty nearly in *statu quo*.

But the disappearance of the meniscus is not the only phenomenon of change of state. A peculiar flickering appearance is noted as indicative of it, together with the formation of striæ, and so characteristic is this feature that it is used by Pictet in some of his most recent work as an indicator of gasefaction.

If a tube is partly filled with a liquid, is sealed and heated, the first indication of a change of state to be looked for is the disappearance of the meniscus. As it vanishes, the flickering striæ appear and a sort of unrest pervades the tube, and quickly the critical state is passed and the liquid has become a gas.

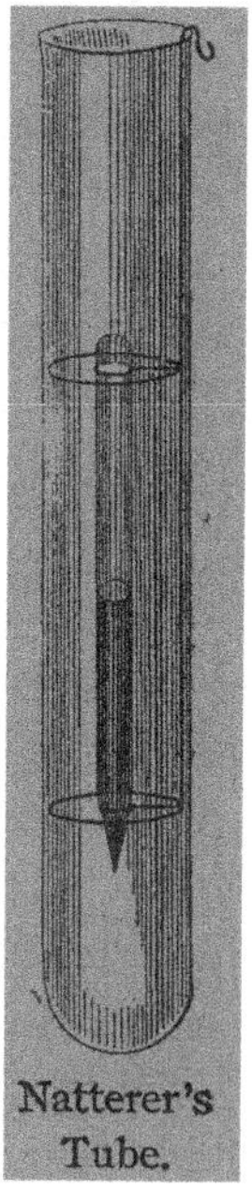

Natterer's Tube.

The phenomenon is conveniently shown in a sealed tube half filled with ether, as shown in the cut. It is mounted within a larger tube filled with paraffin wax. The latter is opaque and solid when cold, but on heating melts and becomes transparent. On heating the wax, the liquid in the inner tube goes through the critical state, the phases can be watched, and the phenomena described above can be seen. If it is to be shown to an audience, the image of the tube is projected upon a screen by the magic lantern, and the phenomena are produced so as to be visible by a roomful of spectators. The sealed tube is termed Natterer's tube.

Hannay and Hogarth, in 1880, in experiments on the critical state of matter, found that several salts, such as potassium iodide and bromide, would dissolve or volatilize in gaseous alcohol at a temperature of 375° C. (707° F.), the whole being contained in a strong sealed tube.

P. Villard (1898) extended the scope of this investigation and got very interesting results with solids and liquids. As a liquid, bromine may be cited. This was placed in a tube with oxygen gas, and the pressure was gradually increased. Normally increase of pressure would be supposed to tend to keep the bromine liquid. But, on the contrary, at two hundred atmospheres, the bromine began to take the gaseous form and to dissolve in the compressed oxygen. The action of the dark brown liquid was exactly that of a substance entering into solution. The gaseous mixture took a darker color at three hundred atmospheres than that of a solution of bromine in water. Villard recalls Cailletet's observation that liquid carbon dioxide dissolves in air. We may also call to mind the *liquide Pictet* (page 170) in this connection.

Bromine is a brown liquid, and is one of the elements; its near neighbor, iodine, is a solid. The latter was found to dissolve in small proportions in oxygen. Formene was another gas which was experimented with. It dissolved ethyl chloride, carbon disulphide, alcohol, camphor, paraffin and iodine. In some cases the gas-solution phenomena were almost reproductions of the critical state phenomena, including the obliteration of the meniscus.

A very interesting suggestion was made by Villard; it was that gaseous solution might take the place of distillation as a laboratory operation.

As the doctrine of the conservation of energy is intimately involved in the liquefaction of air and of all gases, something may be said of the relations of force and energy. This may more appropriately be done as it will bring forward a treatment of the subject which may commend itself to some interested in physics. This treatment of the subject is based on the substitution of two units for three. Usually, force, work and energy are the interrelated units appealed to in treatises on mechanics. The far more desirable way is to follow out the theory of dimensions and to take two of these units only as the foundation stones of the science. These two are force and energy. Work, instead of being awarded an important place, should be treated only as an adjunct and convenient expression of the concrete and accidental. This sounds, perhaps, heterodox. It is really orthodox, and is a move in the direction of avoiding confusion.

As music is built up out of a few notes, as the twenty-odd letters of modern alphabets in a sense are the basic units of the written languages, so we have certain fundamental elements in natural science. These may, for our purposes, be stated as distance or linear space, mass and time. These are familiar to all. The accepted units are the centimeter (0·39 inch), gramme (15·43 grains) and the second. Then there are two derived units, less familiar in their scientific status, and less generally understood, than the others cited above. These are force and energy.

Distance is linear space, space measured along a line, space of one dimension. A foot, an inch, a centimeter, are units of distance. An attempt was made to get an absolute unit by taking one ten-millionth part of the quadrant of the earth as a unit. This is what the French meter was supposed to be, but the measurement was inexact; so the unit is as truly inexact as was the old time barleycorn, except in degree. Its exactness was many times greater, as it approximated at least to a fixed standard, and the length of a barleycorn is as unfixed a standard as could well be imagined, although our system of measures is based on it. Three barleycorns make one inch, and the exceedingly exact standard yard measures carefully preserved by the British and American governments had their origin in the length of a corn of barley. The most recent and scientific unit of length is the wave length of a given monochromatic light. But for everyday purposes the foot is very generally used in this country.

Time is the measure of duration and is the function having a truly international unit, the second. This is an astronomical unit, and might be used as a basis of all others. The proposal to do so has been made, but has never been carried out.

Mass indicates the quantity of matter in a body. It is a somewhat unfortunate unit, as it is constantly confused with weight. A piece of iron has a definite mass, but it weighs one amount at the equator and another amount at the poles. On the surface of the moon it would weigh far less than on the surface of the earth. From one point of view the proper unit of mass would be equal to a pound, or a gramme, or whatever may be taken as the unit of weight divided by the velocity a body acquires in falling through a vacuum for one second. As this last quantity varies at different parts of the earth, it would seem that the unit of mass should in some way be fixed, and that the unit of weight should vary. Accordingly, the quantity of matter in one gramme is taken as the unit of mass. Weight varies, for a pound of sugar at the poles is slightly greater in mass than a pound at the equator. This is very scientific, but does not square with the relative sweetening power of the two pounds.

We have just spoken incidentally of velocity. This is a unit which indicates the distance passed over in a second. As two unitary quantities, time and distance, are involved, it is compound.

We are now ready to see what force and energy are. They are the hardest of all to grasp. Had Faraday and a host of others grasped their significance, the erroneous doctrine of the conservation of "force" would never have been invented.

Force may be variously defined. Newton's definition of it as given by Daniell is "a measurable action upon a body, under which the state of rest of that body, or its state of uniform motion in a straight line, suffers change." But force may be exerted without producing any such change, so that the definition, like many others, is not satisfactory. A copying press applies force to the book it squeezes as long as the screw is left turned down, but it imparts no change of state of motion or of rest to the book. A spring held by a catch of any kind so as to be in a state of tension exerts force against the restraining piece, but there is no question of change of state of motion or of rest. The definition of force as that which exerts a pressure or a pulling stress upon anything, or between any two or more masses, is, for ordinary purposes, an exact enough definition, though not a very elegant one.

The total forces exerted in the universe may vary constantly in amount. There is no such thing as the conservation of force, conservation meaning, in such a connection, constancy or invariability of quantity. Force may be called into existence and annihilated at will. It varies *ad libitum* just as motion does. A man may run or walk or stand still. He thereby creates or annihilates motion. He may do the same for the force he exerts by his own control.

Not many years ago a work was published on the subject of the Conservation of Force. It was made up of extracts from the writings of various scientists which treated of the supposedly true doctrine of the conservation of force. Among other writers Faraday was quoted, and it is curious to see how he could not reconcile the contradictions of the supposed law. He accepted it on the weight of authority of others, his acceptance giving a lesson in humility which some doctrinaires of the present day might profitably study.

All the while the doctrine was an utter falsity and is now discarded absolutely. It is one of the monumental errors of the scientific world. It shows that students of science have their own errors to contend with and guard against. We can reasonably believe, however, that we are not fast bound at present in any such error, at least in the field of physics.

Faraday, who has been cited above, was one of the loveliest figures in modern science and his appearance here is not the only one he makes in the pages of this book, as he appears as one who paved the way for the liquefaction of air and for that of the so-called permanent gases. He it is who gave one of the first blows to this name.

There is one survival of the erroneous doctrine

which, although it only affects the nomenclature, is interesting to notice. It is the term "living force," which cannot be said to have quite disappeared from the language. It was long used as the expression for mechanical energy. The French, who are more conservative than we, adhere to it far more tenaciously, and its equivalent is found in many recent scientific papers in that language. The term is a metaphorical presentation of the idea of force in action, and force in action is nothing more or less than energy. If the action is positive, it is the exertion of energy; if the action is negative, it is the development and consequent absorption of energy.

But the best method of avoiding confusion in modern science is to concentrate the nomenclature and to avoid useless multiplication of terms. So the term living force, picturesque as it is, is very properly abandoned for the more concise term energy.

Energy is a unit which expresses the action of a force along a distance. If a man pushes against a car, and all remains stationary, he exerts, properly speaking, no mechanical energy, but only force. But if the car moves, and he follows, pushing it before him, his force is exerted along a distance, and the compound force-distance unit thus indicated is called energy. Two actions are involved. The man expends energy and gets rid of it. It disappears. But the car receives energy, and in the overcoming of its inertial and frictional resistances an amount of energy is received by it precisely equal to that which has disappeared. This energy is largely converted into heat.

Suppose an athlete holds a dumbbell by his side

and raises it to arm's length. The dumbbell weighing ten pounds and the lift being four feet, he would have expended on it energy represented by the product of force and distance. The force may be popularly expressed in this case as ten pounds, the distance is four feet; the energy expended is forty foot-pounds. The energy which he spent in lifting the dumbbell has disappeared, and in its place has been created the energy now inherent in the lifted mass. By virtue of its position the dumbbell has an ability in recovering its old position to exert energy in its own turn. If the bell drops the four feet, it will, in doing so, lose its favorable position and exert energy. The exerted energy will disappear and cease to exist, but in its place a precisely similar and equal quantity of energy will be developed.

Suppose now that the dumbbell is allowed to fall the four feet through a vacuum. At the end of its fall it will be moving quite rapidly and will be able to strike quite a severe blow. This blow it can inflict by virtue of the energy inherent in it. As this is derived from a fall of four feet, it will be measured by distance and force as before, by forty foot-pounds. If it strikes its blow and comes to rest four feet from its starting point, its energy will disappear, and in some form or other forty foot-pounds of new energy will be created.

The reader will observe that the dumbbell held motionless four feet above its level of rest has the power, when called upon, of exerting in its descent the forty foot-pounds of energy which the athlete exerted on it. It possesses the power of exerting energy, which power is termed potential energy.

Reaching the end of its four-foot fall, it then is charged with energy real and positive, by virtue of which it can inflict a blow. This is the energy of motion or kinetic energy.

Illustrations could be produced in any desired quantity. It would be found that whenever energy disappeared, an equal quantity of other energy appeared. This law holds good always without any exception, and is universally accepted as fixed and invariable. It is most generally expressed by saying that the total energy of the universe is always the same in amount.

It will be noticed that the term "work" has not been used in this brief exposition. Usually, it is one of the first things cited in such cases, and energy is defined as the power of doing work. But it is much better to keep the fact clearly before us that energy is the important and more fundamental unit, and that work is simply another term for development of energy. To "do work" is to expend energy. Our athlete, in raising the dumbbell, expends his own energy, develops new energy, and the latter is the doing of work. The particular energy exerted by the athlete ceases to exist, and is replaced by an exactly equal amount of energy developed in the dumbbell by its change of position. The dumbbell, it would generally be said, has had work done upon it, the lifting of it constituting work; it is far more logical to term this lifting the development of energy in the object acted on.

It would seem somewhat presumptuous to attempt to do away with the term work, and the word is so convenient, and is in such universal use among physicists, that it cannot be dropped. It should, however, be treated rather as a convenience than as a real physical unit, and it should always be understood to be a shorthand term and synonym for development of energy. If work is performed, it is development of energy that is performed, and the object which does the work expends energy in developing new energy.

There is a very simple experiment, which anyone can try, which supplies an excellent illustration of the conversions of energy. An india rubber band is held by the two hands across the mouth, so as just to lie between the lips. It is now stretched. The energy of the experimenter is spent on stretching the band; some other equivalent of energy must be developed to take its place. As the band stretches, the lips can feel it grow warmer. The mechanical energy expended in stretching it is converted into the kinetic energy of heat. It is allowed to resume its original length. In doing so, it exerts energy. It has only the kinetic energy of its heat to call upon. Accordingly, it grows cool as it resumes its original length, and the lips feel the cooling effect. It illustrates the law of the conservation of energy excellently, and is particularly interesting to the reader, as it applies very strikingly to the expansion and contraction of gases.

We can now appreciate the conception of a reservoir of energy. The pound weight, held at four feet elevation, exerts no energy, but does exert force. It is a reservoir of energy in potential form. The same weight, moving with the velocity acquired by a fall of four feet, is a reservoir of energy in kinetic form. Brought to rest after its fall, the kinetic energy it was charged with disappears and it is no longer a source or reservoir of energy.

When energy is expended by any mechanism, the new energy developed to replace the old in the world's scheme, and to keep the amount of the world's energy invariable, is apt to take largely the form of heat energy. A railroad train has expended on it the energy of the locomotive. Suppose it runs a mile upon a dead level. At the end of the mile it occupies a position not one whit more advantageous than when it started, as far as energy of position is concerned. Yet the fire in the fire-box of the engine has fiercely burned over the mile run, and the energy of the sun of bygone ages, stored up for geologic epochs in the inert coal, has been expended. What energy has been developed to take its place and keep up the balance?

It is energy of heat. The wheels have pounded over the rails, heating themselves and the rails, their journals and the journal-boxes have been heated, and even the energy expended on overcoming the air resistance has heated it a little, and the sides of the cars have been heated a little also. This heat is absolutely useless, or even pernicious. We cannot move a train along a level roadbed, we cannot drive a ship across the level plain of the ocean, without expending energy which we can never recover. It goes into the storehouse of nature, never to be recovered by man until another great step in advance is made. The liquefaction of air has in it a germ, dimly recognizable, which may enable us to utilize the low forms of energy with which nature is charged. The ocean path, and the steamer which traverses it, at the end of the Atlantic trip may have received one hundred and forty thousand horse power days of energy. Now it is all lost to man. Man's ingenuity perpetrates no more wasteful and unsatisfactory acts than the transfer of himself and his possessions across the ocean or over continents. The thirty thousand horse power engines of the transatlantic liner are no more a triumph of human ingenuity than in their enormous wastefulness of practically one hundred per cent. they are a concession to his inability to utilize the energy of the universe.

This brings us face to face with the doctrine of entropy. We have seen that the low degrees of heating produced by the friction of machinery, and which represent its wasteful resistance, are lost forever to us. The potential chemical energy represented by the separation of carbon and oxygen is the energy of carbon or coal which can be burned under a boiler when it unites with the oxygen of the air. This is one of the world's energies which can be utilized by man, and these energies are called available energy or entropy. The world's coal is being burned up, its forests are being destroyed, machinery is adding to the irreclaimable energy of the world, and, by the doctrine of the conservation of energy, is destroying that same quantity of available energy; hence the entropy of the universe is becoming smaller day by day.

Clerk Maxwell saw the possibilities of the utilization of the unavailable energies of the universe. It is provoking to know that our great ocean of air is pulsating with molecular energy which we do not utilize. Yet we do utilize it in a sense in compressed air motors, we call upon it in liquid air work, but Clerk Maxwell's dream of the utilization of the lost energies of the universe will not come true by the application of liquid air and liquefied gases to motors.

A popular paradox, which has been much discussed, may be used to give an example of the doing of work at the expense of the low grade heat of the air and of other matter. A steel spring is placed in tension or is wound up. It is then dissolved in acid. The question is, What becomes of the energy which seems to be present in the spring, and ready for utilization? One theory is that there is present in it no energy which in any way is due to its being wound up. When first wound, the energy expended in the operation develops new low-grade heat energy, and the spring is slightly heated. Then it loses the heat in a few seconds, and there is no longer any more energy in it wound than unwound. Therefore, it dissolves in acid without having any special energy to account for.

Now, the question may be asked, How can the spring, if it has no energy, drive a clock? It does this, not at the cost of any mechanical energy due to its tension, but utilizes the low-grade heat energy of which we have been speaking. As it drives the clock it gets cool, and the energy required to drive the clock is represented by this cooling. As air circulates around it, it recovers immediately any loss of temperature, so that no loss of heat is practically discernible. But the clock is driven primarily by the heat of the air, by heat such as is usually treated as unavailable. The india rubber band experiment described on page 32 is an exact illustration of the point involved.

Elsewhere the possibility of using liquid air as a substance for the storage of power is alluded to. If this were done, an engine could be driven by it exactly as by steam, except that the heat would be drawn from the atmosphere instead of from a burning furnace of coal, and there would be a utilization of low heat energy.

CHAPTER II.

HEAT.

Heat and its measurement—Thermometers—The zero point—The Celsius or Centigrade thermometer scale—Fahrenheit's thermometer scale—The absolute zero—Its basis—Coefficient of expansion of gases—Determination of temperatures in the liquefaction of gases—Different liquids used in filling thermometers—The air thermometer—The hydrogen thermometer—Details of its construction—Electrolytic hydrogen—The hydrogen or air thermometer formula—The thermo-electric thermometer—Onnes' instrument and details of its construction—Its calibration—The electric resistance thermometer—Calorimetric determination of temperatures.

Heat has been referred to. While all have a general idea of heat, the basis of the different thermometer scales may be spoken of, and the absolute zero defined more fully.

Various thermometer scales have been proposed, and three are in general use. Thermometers generally indicate the temperature by the movements of an indicator over a graduated scale. Mercury and colored alcohol are the substances whose expansion by heat is utilized for ordinary thermometers, and the upper surface of the column of mercury or alcohol forms the indicator. The scales had to be divided on some system or other. The first thing to be settled was where to place the zero point at which to begin the division. Fahrenheit placed it well below the freezing point of water. Reaumur and Celsius placed it at the point where ice melts, which is the freezing point of water also. A name for this point is required, and the name zero, of Italian origin, from the same Arabic root as our word cipher, is given to it. Zero seems to apply more to thermometric scales than to others, simply because we are more familiar with this class of scales than with hydrometers and other scale-bearing instruments.

At the zeros of the above thermometric scales an active molecular motion exists; there is a quantity of heat present in all things, at and far below the zeros; ice is hot, ice water is hot, frozen mercury is hot. This seems illogical; nothingness on the thermometer scale should indicate nothingness of heat. As thermometer scales are graduated now, their zero points are placed in a locus of very considerable heat. They can only be called points of relative cold; we think them cold because of our physiological peculiarities. Bacteria do not seem to think that ice is cold; at least they live through freezing unimpaired in vitality.

Two easily produced temperatures are used for establishing thermometer scales. One is the boiling point of water, the other the melting point of ice. By comparatively simple apparatus these temperatures can be reproduced at will, without need of the application of any difficult correction. For the graduation of ordinary thermometers no correction is applied, although the barometer reading should be taken into consideration.

The standard scientific thermometer is the Celsius

or Centigrade instrument. In this the temperature of melting ice is taken as zero, that of boiling water, or, more accurately, of steam at atmospheric pressure, as one hundred, and the space between and above and below these points is uniformly divided off on that basis.

One account says that Fahrenheit attempted to get absolute cold, that he made a freezing mixture with ice water and salt, or sal ammoniac, and took its temperature as being perfect cold. Then he took the temperature of the human body as another datum point, and tried to have the freezing point of water one-third way between his zero and the human body temperature. Of the three devisers of thermometric scales, he was the only one who made an attempt to get a genuine zero. In the early days of the eighteenth century, when Fahrenheit was doing his work, the kinetic theory of heat, which is what we are here describing, had not been evolved. It was in 1724 that his low temperature experiment was published.

Another explanation of Fahrenheit's thermometer is that he took as his zero a temperature observed at Dantzig, Prussia, which he found that he could always reproduce by salt and ice. He computed that at that temperature, which he believed to be the absolute zero, as he interpreted it, his thermometer contained 11,124 parts of mercury, which expanded to 11,156 parts in melting snow. This gave him 32 parts expansion, or 32 degrees. In boiling water he found his mercury had increased to 11,336 parts. This gave him (11,336—11,124=212) 212 parts or 212 degrees between his zero and the boiling point.

Absolute cold has been defined. It is the temperature at which all heat energy ceases—when the molecules would cease to vibrate, when molecular death would occur. This point is the starting point of the theoretically correct themometer scale—its zero. Were it not too late, the thermometer scales of the world should be based on this point as a starting point.

This point is termed the absolute zero. It lies at 273° C. below the Centigrade zero (—459·4° F.)

A good temperature for a living room is 20° C. (68° F.) It would on the absolute thermometer be 273+20=293° C. (527·4° F.) Instead of complaining that the mercury has gone up to 99° in the shade, we might correctly call it 558° in the shade and feel that we had better ground for complaint. The absolute zero has had a definite place assigned it, based on the properties of the form of matter which is acted on by heat with perfect freedom. It is the form of matter in which the molecules are free to move under the influence of heat unhampered by any individual attraction, in other words, the gaseous form of matter.

Imagine a quantity of gas which we will suppose to have, at the freezing point, a volume of 273 cubic inches. If we heat it 1 degree Centigrade, it will become 274 cubic inches. Another degree rise of temperature will make it 275 cubic inches, and so on. If we cool it 1 degree Centigrade below the freezing point, it will become 272 cubic inches, and so on all the way down. The paths of vibration of the molecules thus grow smaller and smaller with each reduction in temperature, until we are led to the conclusion that, when the temperature has been lowered 273 degrees, the gas, losing 1 cubic inch at each degree reduction, will have lost its entire volume, or will have been reduced as near to a volume of nothingness as it can get. Now, the idea of its having a volume of nothingness or of a gas losing its entire volume being absurd, we substitute the theory that, at 273 degrees below freezing, the paths of vibration of the molecules will become infinitely short, that their length will become nothing, and that the molecules will rest.

The absolute zero is based on these considerations. The proposition is stated and proved above in a very crude way, but it gives a simpler presentation of the subject than is given in the ordinary statement of the subject. The law of the expansion of gases by heat may be thus more scientifically stated.

If we start with a volume of gas at any temperature and apply heat, it will increase in volume. For equal increments of heat it will increase identical amounts, or for equal increments it will increase equal portions of the original volume. Confining ourselves now to the Centigrade scale, we find that for increments of temperature of 1 degree, the volume of a gas will increase by $\frac{1}{273}$ of what its volume would be at the temperature of melting ice or zero Centigrade. This is termed the coefficient of expansion of gases. The same occurs for reductions of temperature. Therefore, at 273° below zero no more reduction in volume will be possible. At this point the motions of the molecules must stop—it is absolute zero.

The determination of the low temperatures employed in experiments on the liquefaction of gases is naturally attended with difficulty. The mercurial thermometer had to be discarded because the metal solidified at a comparatively high temperature when referred to the degree of cold attained in the experiments. Even in Faraday's experiments the mercurial thermometer was discarded in favor of the alcohol thermometer. The degrees on the instrument he employed, which was a Fahrenheit thermometer, were graduated below 32° F. into degrees respectively equal in length to those between 32° F. and 212° F. on its scale. He got down to —110° C. (—166° F.) Not reaching the critical temperature of oxygen, he naturally failed in liquefying it. What Wroblewski and Olszewski term "a dazzling demonstration" (eine glänzende Bestätigung) is given by an experiment of Natterer, who shows that the incredible pressure of 3,000 atmospheres alone is insufficient to liquefy oxygen. When it is realized that the pressure in a modern cannon at its maximum is about two-thirds of this amount, it can be seen what the scope of Natterer's experiment was.

Natterer used a thermometer filled with phosphorous chloride, as he orally informed Wroblewski or Olszewski (*Wiedemann's Annalen,* 1883), and Cailletet, in his work on low temperatures, used a carbon bisulphide thermometer. Wroblewski and Olszewski used a hydrogen thermometer constructed on the model of Joly's air thermometer (*Poggendorff's Annalen,* 1874).

Wroblewski and Olszewski found a slight discrepancy between the readings of a carbon bisulphide and a hydrogen thermometer. The carbon bisulphide instrument read about 2 degrees Centigrade lower than did the hydrogen thermometer. This reading was but a few degrees above the solidification point of carbon bisulphide, and under such conditions, namely, an approach to its solidification temperature, an irregularity in expansion and contraction is always to be looked for in a liquid. The carbon bisulphide thermometer scale is graduated on the basis of higher temperatures—the coefficient of expansion is much greater near the solidification point than it is higher up the scale.

The same observers note that when the carbon bisulphide freezes in the thermometer, the tube breaks into several pieces. They found that a couple of minutes' evaporation of ethylene in a vacuum was sufficient to freeze bisulphide of carbon. They put its freezing point at about —115° C. (—177° F.) It melts, they state, at about —110° C. (—166 F.) Common 95 per cent. alcohol thickened at —129° C. (—200·2° F.) and froze solid at about —130·5° C. (—203° F.) Methyl alcohol (wood alcohol) was easier to freeze than ordinary alcohol. Phosphorous chloride froze at about —111·8° C. (—169° F.) These substances, it is claimed, were never frozen before this period (*Wiedemann's Annalen*, 1883).

The figures show that these liquids are not available for low temperature thermometers, and are cited here for the purpose of showing that fact.

The ordinary mercury and spirit thermometers, familiar to all, and their modifications, the carbon bisulphide and other themometers of liquid contents, then, are useless for very high or very low temperatures, their liquid contents volatilizing or freezing

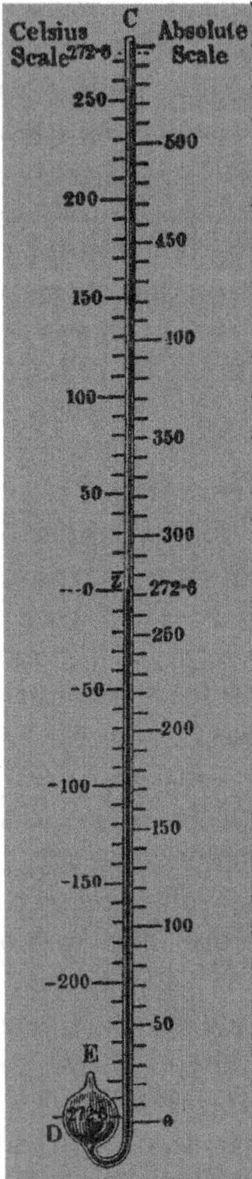

Gas Thermometer of Varying Volume.

solid at high and low temperatures respectively. Air was substituted for the liquids, and thermometers operating by its expansion when heated were devised. The cut shows the general features of construction of one of these. The bulb contains air at *E*. Mercury, *D*, lies in the tube, cutting off the end from the bulb. As the air expands, it forces the mercury up; as it contracts, the mercury descends. This is a thermometer of changing volume. It is not so satisfactory as the air thermometer of constant volume.

The cut also shows the relation of the Centigrade and absolute thermometer scales. On the left is engraved the Centigrade or Celsius scale, with its zero marked 0 at the point of melting ice, its 100° mark at the point of boiling water, and —273° at the absolute zero. On the right is the absolute scale, on which ice melts at 273° and water boils at 373°.

There is a third thermometer scale which may be mentioned here, although it is rarely used in scientific work; it is called the Reaumur. The zero is the same as the Centigrade zero, and the boiling point is made to read 80°. This is the basis for its expansion up and down. At the absolute zero its reading is —218·4°.

If, as the temperature changes, a confined gas is kept at a constant volume, its pressure will vary; it will rise as the temperature rises and will fall as it falls. If we provide a means for measuring the presure of the confined gas, we can determine therefrom its temperature.

The word gas has been used instead of air, for other gases can be used with equal accuracy. For the extraordinarily low temperatures encountered in gas liquefaction investigations an air thermometer is useless, because the air liquefies. Just as mercury gave place to alcohol in liquid thermometers for low temperature work, so did air give place to hydrogen in gas thermometers.

The constant volume hydrogen thermometer as a standard temperature-determining instrument for low temperature work is of simple construction, based on the phenomena of change of pressure under change of temperature in a gas kept at constant volume. This is the converse of the expansion and contraction of matter when heated. It is practically only applicable to matter in the gaseous state.

If a thermometer of the ordinary construction is heated until the tube is filled to the top by the expanding mercury or alcohol, a little more heat will crack the glass, and the contents will escape. The expansion of liquids when heated generates enormous pressures. But if the thermometer were filled with air or hydrogen or other gas, it could be heated very hot, probably to the melting point of the glass, before it would give way.

In the mercurial, alcoholic or other thermometer with liquid contents, the heat is measured by the expansion of the liquid, which is purposely so placed as to be perfectly free to expand. In the air, hydrogen or other gas-filled thermometer of the type we describe, the gas is kept at constant volume, and the pressure it exerts is measured. A diagrammatic representation of the construction is given, which can be readily followed by the reader.

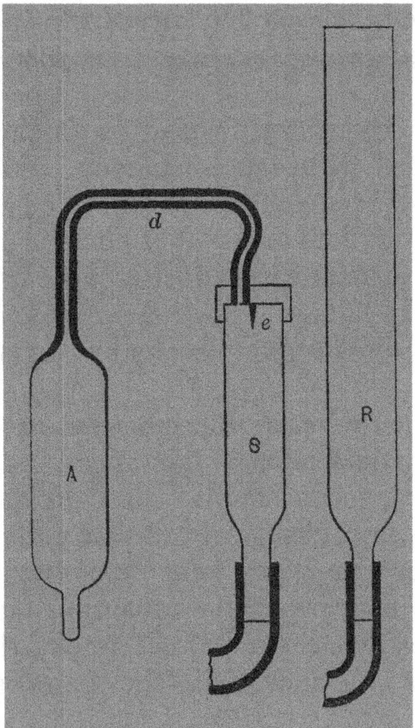

Details of Hydrogen Thermometer.

A bulb, *A*, is filled with perfectly dry pure hydrogen. From its top a capillary tube, *d*, rises and connects with a mercury tube, *S*. The connection is preferably so made that the top of the mercury tube shall be perfectly flat. The capillary tube, *d*, enters a little to one side of the flat top of the tube, *S*. In its center a point, *c*, of glass, ivory, steel, or some material unattacked by mercury, is attached, which points downward.

The bottom of the mercury tube is reduced in diameter, is open, and an india rubber tube has its end thrust over it. The other end of the india rubber tube is connected to the bottom of another glass tube, *R*, termed the manometer tube. When the apparatus is set up, this tube can be moved vertically up and down. A clip moving up and down a vertical rod on a firm stand and attached to the tube enables this to be done. The tubes, *R* and *S*, contain mercury.

If the tube, *R*, is raised or lowered to the proper point, the mercury in *S* can be brought to precisely the level of the point. This is indication by a point, a very delicate means of fixing the level of mercury. It is used in barometers in adjusting the level of the mercury in the cistern, and is taken as being sensitive to one-thousandth of an inch. The mercury as it rises reflects, mirror-like, the point. When the latter touches the mercury, the point and its reflection form a continuous line. If the mercury is raised too much, a dimple forms on its surface. The appearance is unmistakable.

By the manipulation of the observer sliding the manometer up and down the rod, the mercury is brought into accurate contact with the point, *e*. This is done for every reading of a temperature. This being the case, it is obvious that the heights of the upper surface of the mercury in *R* will vary according to the pressure of the gas in *A*. As this is greater, the surface of the mercury in *R* will be higher; as the pressure is less, the level in *R* will be lower; the readings being taken only when the mercury in *S* has been brought to its exact level by raising or lowering the manometer tube, *R*. The greater pressures correspond to greater heat of the contents of the bulb, *A*, the lesser pressures to lower heat. By measuring the difference of level of the surfaces of mercury, the data for calculating the heat are given.

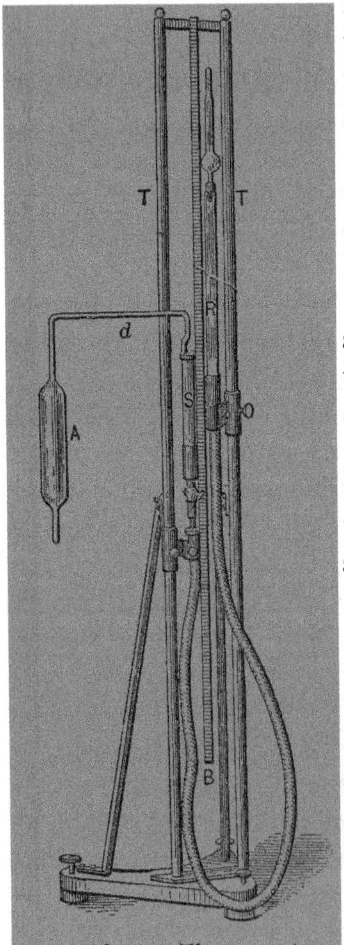

Hydrogen Thermometer.

The height is best read by a cathetometer. This is a telescope with cross wires across its tube, in the focal plane, and mounted to be moved up and down a vertical rod on another stand, without ever departing from a perfectly horizontal position. A vertical scale of great accuracy of division is mounted near the manometer tube. The telescope is focused from a distance upon the apparatus. The mercury is adjusted by moving the manometer tube until the mercury touches the point, *e*. The telescope is slid up and down until the image of the surface of the mercury in the manometer tube, *R*, exactly coincides with the cross-wire as seen in the telescope. The telescope is now swung in a horizontal arc if necessary, until it takes the vertical scale into the field. The reading of the scale gives the height of the mercury. The same is done for the mercury in the tube, *S*; the difference gives the pressure of the hydrogen in units of a column of mercury.

As the point, *c*, is supposed never to change position, the scale may be adjusted so that its zero is at the level of the point. For a series of readings one reading of the point level would in any case suffice.

The general mounting and disposition of parts of a constant volume gas thermometer are shown in the cut. *A* is the gas bulb, *d* the capillary tube, *S* the mercury tube, *R* the manometer, *T T* the frame, and *B* the vertical scale. Clamps are arranged to slide up and down the side rods of the frame so as to adjust the levels of the mercury vessel and manometer tube.

Prof. H. Kamerlingh-Onnes, of Leyden, prepares hydrogen for his hydrogen thermometer by electrolysis as described in the most general terms on page 148. A very carefully constructed apparatus is used for the purpose. The interior of the hydrogen bulb and tubes are most elaborately cleaned with chemical solutions and distilled water and dried before the introduction of the hydrogen, and various modifications have been introduced by him.

At the risk of trenching upon the determination to avoid the introduction of much mathematics into this volume, the very simple calculation used in reducing the hydrogen thermometer readings to the standard is given. The reader may be assured that it is not as complicated as it appears.

To obtain the formula for the thermometer, the bulb is immersed in melting ice or snow, and the manometer is adjusted so that the level of the mercury in *S* just reaches the point, *c*. (See cut on page 46.) The readings of the heights of the two mercury columns are now taken.

The calculation is based upon equating two expressions for the weight of hydrogen contained under the conditions of the two readings in the bulb. Let S_0 be the specific gravity of the gas in the bulb, let V_0 be the volume of the bulb, and v_0 the volume of the capillary tube; let H' be the height of mercury column, measured from the fixed level of the point, *c*, to the level of the upper surface of the mercury in the manometer tube increased by the height of the barometric column. S_0 is taken at 0° C. and 760 mm. barometer. The weight then will be expressed by

$$S_0 \left(V_0 + v_0 \right) \frac{H'}{760}$$

Next the bulb is placed in the substance whose temperature is to be determined. Let *k* be the coefficient of expansion of hydrogen (0·00367), *a* that of glass, *t* the temperature to be found, and H the new difference of levels of mercury columns increased by the height of the barometric column. The weight of hydrogen, the same as before, is

$$S_0 \left(V_0 \frac{1 + a\,t}{1 + k\,t} + v_0 \right) \frac{H}{760}$$

And equating we have:

$$S_0 \left(V_0 + v_0 \right) \frac{H'}{760} = S_0 \left(V_0 \frac{1 + a\,t}{1 + k\,t} + v_0 \right) \frac{H}{760}$$

Solving these with respect to t, we find that—

$$t = \frac{(V_0 + v_0)\,(H - H')}{V_0\,(k\,H' - a\,H) - v_0\,k\,(H - H')}$$

This seems rather a complicated formula, but the use of the hydrogen thermometer is amply justified by the sensitiveness of the instrument, its great accuracy and great range. It can be used from the temperature of liquefied gases up to that of the melting point of glass.

If two dissimilar substances have their ends connected so as to make a circuit, and if both are conductors of electricity, a current of electricity will pass through them as long as one of the contact points of the dissimilar substances is hotter or colder than the other. The effect is termed thermo-electric and the junction is termed a thermo-electric junction. The current with a single pair of junctions will be due to a very slight potential difference. The greater the difference of temperature, the greater will the potential difference be. If means are provided for measuring the potential difference, and if the temperature of one of the junctions is known, then the amount of the potential difference will give data for calculating the temperature of the other junction.

The thermo-electric junction has been much used in low temperature work. The conductors may be varied a good deal. A standard type is German silver—copper. The former metal is an alloy of

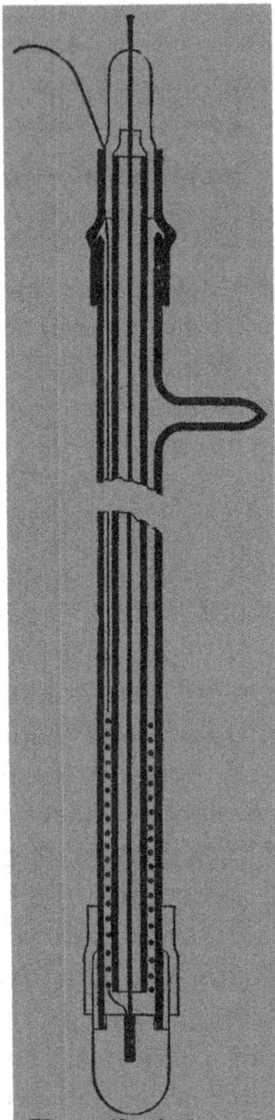
Kamerlingh-Onnes' Thermo-electric Thermometer.

copper, nickel and zinc. Other couples are German silver—copper sulphide (Becquerel's); German silver—zinc-antimony alloy (Noë's); iron—bismuth-antimony alloy (Clamond's).

The ordinary practical unit of potential difference in electric work is the volt. In the thermo-electric junction the difference is so slight that it is usually measured by micro-volts, or millionths of a volt. The measurement of the potential difference is effected by means of a sensitive galvanometer. It is unnecessary to give the details of this operation.

As an example of the thermo-electric couple, as applied to the determination of low temperatures as encountered in the liquefaction of gases, an illustration of the couple used in the cryogenic laboratory of the University of Leyden is given. This laboratory, specially fitted with elaborate apparatus of the Pictet type, has won considerable fame, and, under the charge of Prof. H. Kamerlingh-Onnes, much excellent work has been done there. In a journal recently started in Berlin, and which is devoted to the topic of compressed and liquefied gases (*Zeitschrift fuer comprimirte und fluessige Gase*), is given a description of the principal apparatus in the laboratory, which may be advantageously studied by those specially interested in the liquefaction of gases.

The cut gives the section of the thermo-electric couple. It is formed of a straight German silver wire soldered at its lower end to a thin copper wire. The latter is coiled into a helix.

The cut shows in the center the German silver wire as a straight black line. It lies within a glass tube. Around the outside of the latter is wound a thin silk-covered copper wire. The ends of the two are inserted into a block of copper and soldered The silk insulation serves to keep the copper wire from touching itself in its successive turns. Another way of arranging it is to melt and wind a thin glass filament around the tube and wind the wire in the grooves it forms.

Outside of the inner tube and of its winding of copper wire is a second glass tube. By india rubber tubing the junctions are completed as shown.

The copper block at the bottom is turned off to a shoulder, so as to fit inside the outer glass tube. A thin tinned sleeve of copper is soldered to it, and this sleeve goes outside the lower end of the outer glass tube. The joint is made good with melted sulphur. By the side branch the apparatus is filled with dry air, two apparatus being joined by a rubber tube for the purpose.

By immersing the copper block in anything colder or hotter than the wires themselves are, a temperature difference is established. One of the junctions of two dissimilar metals is at a temperature different from that of the rest of the wires and of the other junction. If the ends of the wires are connected in circuit with a galvanometer, it will be deflected by the current due to the thermo-electric effect.

Such an instrument is calibrated by comparison with an air or hydrogen thermometer, and indicates changes of heat with great delicacy. A moment's reflection will show that where two dissimilar metallic or other conductors are joined, so as to form a circuit, there will be two junctions of dissimilar conductors; the circuit must include two thermo-electric junctions. The general law is that the electromotive force developed by a thermo-electric couple varies with the excess or depression of temperature of one junction over that of the other junction, which must lie in the rest of the circuit. This law holds measurably true for excessive variations. For a German silver—copper couple, the potential difference is about one hundred-thousandth (0·00001) of one volt per degree Centigrade, or five-ninths of this amount per degree Fahrenheit.

Many substances possess the property of opening a path through the luminiferous ether for electricity. A constant discharge at very low potential can occur through such a path. The discharge of electricity is called a current, the substance whose presence opens the path is termed a conductor. Copper wire is one of the best conductors known, and is very familiar in such application. House work for telephones, electric lights and electric bells is generally, almost universally, done with copper wire. It is rapidly being introduced on main telegraph and long distance telephone lines.

Electric conductors, like water pipes, may be good or bad conductors. A smooth-lined water pipe will carry or conduct more water than one with rough interior. Some metals will conduct electricity better than others. A metal of poor conducting power is said to have great or high resistance. Iron is of rather high resistance, platinum is of rather high resistance, copper and silver are of low resistance.

The same conductor varies in resistance with its temperature. Generally, the hotter it is, the higher is its resistance, and the colder it is, the lower is its resistance. It is believed that at the absolute zero of temperature, the resistance of copper or of iron would be abolished almost entirely or even entirely. Then the thinnest wire could conduct the horse power of Niagara to any distance without loss.

Based on the above facts, the platinum wire resistance thermometer is constructed, and while it is also an instrument adapted for high temperatures, it has been used with the best results in the investigation of the low temperatures encountered in the investigations of liquid air and liquefied gases.

Olszewski in an article in the *Philosophical Magazine* for 1895, claims that his associate, Witowski, was the first to successfully use the platinum resistance thermometer for the determination of liquefied gas temperatures. In its usual form it is very simple, such simplicity being possible because liquid air and the liquefied gases in which it is used are excellent insulators. As the wire is to be surrounded by them, the fact that it can be immersed uninsulated without short-circuiting conduces to simplicity of construction and to sensitiveness.

The principle of construction can be seen in the cut, in which is given a representation of an apparatus used by Prof. Dewar to show the decrease of resistance of a wire when the temperature is lowered. The tube is a vacuum tube containing liquefied oxygen or liquid air. In it is immersed a coil of fine platinum wire, held in shape by a sheet of mica with notched edges, around which it is wound. Two heavy platinum wires serve as connectors. These are so large in diameter, and so short, that their resistance may be regarded as quite negligible. The wire with the mica sheet and its mounting is the thermometer.

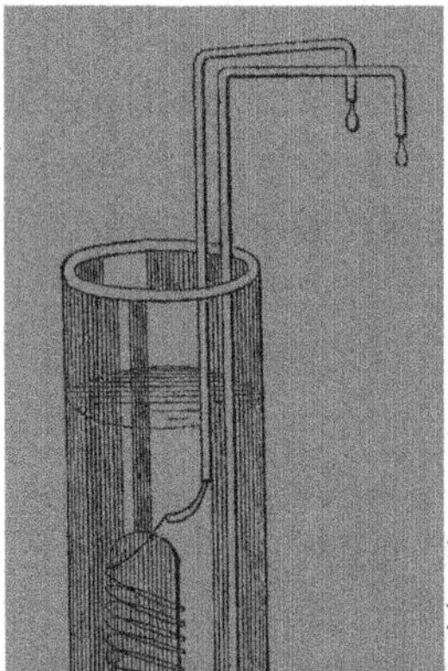

Principle of the Electric Resistance Thermometer.

Another form of construction provides for a more thorough exposure of the platinum wire to the changes of temperature by separating it as far as possible from contact with other matter than the liquefied gas. Out of very thin mica or ebonite a frame is made whose cross-section is a sort of hexagonal star. Around this the platinum wire is wound. This arrangement provides a coil of wire in contact with a non-conducting substance only at a comparatively small number of points, six for each complete turn of the coil. It is a disposition of the wire which secures a considerable length in a small space, and which leaves the wire free to be in most intimate contact with the material surrounding it. The temperature of the wire changes with the greatest quickness, and the thermometer is of the most sensitive type yet devised. It is due to Prof. Olszewski.

The platinum wire he employed was 0·025 millimeter, or about one-thousandth of an inch in diameter. The successive turns of the wire were one-half to one millimeter, or one-fiftieth to one twenty-fifth of an inch distant from each other.

Witowski's electric resistance hermometer was constructed with a view to keeping the platinum wire out of contact with the liquid it was to be immersed in. The wire was wound upon a copper cylinder with mica insulation. It was inclosed in a copper foil cylinder, and was hermetically sealed therein.

Callendar and Griffiths studied the subject of platinum resistance thermometers in the Cavendish Laboratory, at the University of Cambridge, England. They reached the conclusion that the instrument is accurate to one-thousandth of a degree change of temperature. This fact, together with its great sensitiveness, makes it an ideal instrument for use with non-conducting liquids such as liquid air.

The thermometers are used by passing an almost infinitesimally small current through them and accurately measuring the resistance. It varies in degree with the temperature, and the instrument may be standardized by the hydrogen thermometer.

Finally, there is one more way of determining

what may be termed extreme temperatures, which was tested by Cailletet in some of his recent work, which showed that it was reliable for liquefied gas temperatures. A piece of metal of known weight and specific heat is immersed in the liquid whose temperature is to be determined. After it has attained the temperature, in five minutes, more or less, it is removed and transferred to a calorimeter or apparatus for determining the quantity of heat in it. The simplest calorimeter is a vessel of water, and for rough work this can be used. The piece of metal is quickly thrown into a vessel of water of known weight and temperature. The change of temperature of the water brought about by the introduction of the piece of metal, by a simple calculation gives the temperature of the piece of metal.

For scientific work some of the more accurate forms of calorimeter are used, which it is unnecessary to describe here. The calorimeter method has been very rarely used, and is only mentioned here on account of Cailletet's paper of 1888.

CHAPTER III.

HEAT AND GASES.

The perfect gas—The ultra-perfect gas—Energy expended in heating a gas—Specific heat at constant pressure and at constant volume—Atomic heats and variations of same from equality with each other—Adiabatic and isothermic expansion of gases—Carnot's cycle—The perfect heat engine—Available and unavailable energy—Unavailable energy rendered available by liquid air—Latent heat of melting, of vaporization, of expansion—Boiling a cooling process—Expansion a cooling process—The spheroidal state—The Crookes layer—Experiments and illustrations—Utilization of the spheroidal state in low temperature work and in liquid air investigations.

The perfect gas has certain defined characteristics, or it may more properly be said, should have them; for a perfect gas is a rarity, and some of the representative methods of liquefying air are supposed to be based on the fact that air is not a perfect gas.

If a gas is compressed, energy is expended upon it and an equal amount of energy is developed in the gas. This appears largely and principally as heat. Were air a perfect gas, it would all appear as heat. But in the case of air at 19 atmospheres, about $\frac{1}{63}$ of the energy spent in compressing it fails to show itself as heat energy.

Following this out, a perfect gas expanding against pressure and developing energy should lose heat exactly equal to the energy it expends in the expansion. But here, too, there is a loss of heat energy. The expanded air is a little cooler than it ought to be, because the act of expansion requires energy to be spent upon the molecules in some not well understood way. Hence there is a greater cooling than would be indicated by the energy expended.

Hydrogen is a gas that acts in the opposite way. It requires more energy to compress it than would

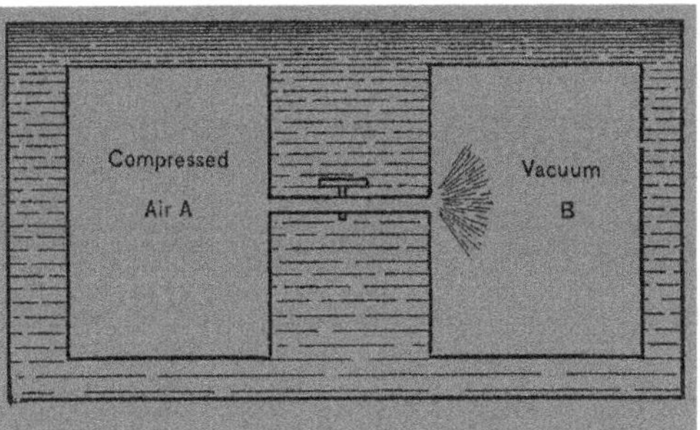

Joule's Experiment.

be indicated by the heat developed, and in its expansion it does not get as cool as it ought to. Hence it is a more than perfect gas—an ultra-perfect gas.

There is no perfect gas known. None has ever been found capable of standing the tests which a perfect gas should respond to.

One test to which a perfect gas should respond is the following: Two gas receptacles are connected by a tube. One is charged with gas under pressure, the other has a vacuum produced within it. A cock upon the pipe connecting them is closed so as to maintain the condition described. The two connected vessels are immersed in water and all is left standing until the gas receptacles, the gas in one of them and the water surrounding them, are of even temperature. Now the gas cock is opened.

The compressed gas streams out of the one receptacle into the other. As it expands it exerts mechanical energy. This must be supplied from some source, and heat energy is called upon. The expanding gas grows cooler. The gas in the other vessel is compressed. Energy is developed and must show itself; it appears as heat. The gas in the second vessel is heated.

If the gas were a perfect gas, the heating would exactly balance the cooling, and the temperature of the water would be unchanged. Joule tried the experiment, and thought that the temperature of the water was unchanged. There was so little alteration that it completely escaped recognition; a thermometer with bulb immersed in the water was apparently unaffected. But there was a difference. If air was used, the temperature would fall, and the same is to be said for most other gases.

These differences are so slight that it is only by delicate tests that they can be detected. The scientific incredulity of Joule and Thomson led them to try a simple experiment, which may be described here.

A tube is provided with an air-tight piston. A diaphragm extends across its center. This diaphragm is made of a porous material which will only permit the passage of air with some difficulty. If the piston is forced in, the air is compressed and heated. It escapes through the pores in the piston and expands as it escapes. Now, as the expansion of the air exactly undoes the compression, there should be an exact balance of energy expended on the air on the piston side and energy expended by the air on the free side. Hence, the escaping air should be of the temperature of the atmosphere. But it is found to be lower. Air is an imperfect gas. If hydrogen be substituted for air, the temperature is higher. Therefore, hydrogen is an ultra-perfect gas.

Our ancestors had their own way of looking at gases. They attempted to classify them into permanent and non-permanent gases, for they believed that there were some gases which could not by any degree of cold or pressure be liquefied or solidified. These they called permanent gases. Then there was adopted a rather crude division of gases and vapors. The latter were easily reducible to the liquid form, the former were not. This was profoundly unscientific and inexact. It left it largely a matter of fancy when a gas should be called a vapor. It led to confusion of ideas, and such expressions as vapor density, tension of aqueous vapor, and the like have done much to obscure the student's view of the status of things. But it is uncertain when the terms which have more or less had this effect will have better ones substituted for them. Perhaps it

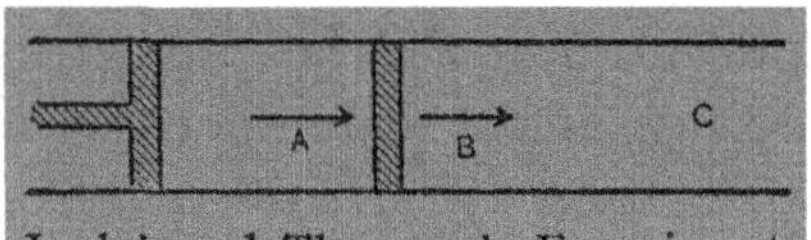

Joule's and Thomson's Experiment.

may be that it will be hard to find better expressions.

The best definition of a vapor is, perhaps, the following: A gas which, by the least increase of pressure or reduction of temperature, would be reduced, in part, to a liquid. The term vapor, thus defined, is subjective. If a liquid is introduced into a vacuum, it evaporates in whole or in part. If enough is introduced, an excess of liquid may be left, and will lie on the bottom of the vacuum chamber. The gas filling the chamber is then a typical example of a vapor as thus defined.

In the upper part of a barometer tube there is present volatilized mercury, mercury gas or vapor, in exceedingly small amount. This varies in amount with every change of temperature and of barometric pressure. In the outer chamber of Dewar's bulbs for holding liquid air a globule of mercury is seen. This fills the vacuous space with mercury gas or vapor. These are examples of vapor as defined above.

In the present work an endeavor will be made to adhere to the term gas, to the exclusion, as far as possible, of the term "vapor." As we have little to do with chemistry, the subject of gases will be compar tively simple, as they will be dealt with as physical concepts. Thus, although air is a mixture of two principal gases, oxygen and nitrogen, with smaller amounts of others, such as argon, gaseous water, and carbonic acid gas, it will be spoken of as a gas when only its physical relations are under consideration.

Another definition of vapor is, a gas at any temperature below its critical one. It is a gas which by pressure alone can be reduced to the liquid state.

However little one may fancy the term vapor, owing to the varied definitions given for it, there are some cases when its use is almost obligatory. Water vapor is one of these. If we speak of water gas, it is taken to indicate a combustible gas, containing free hydrogen, but no water, which is produced by passing steam through incandescent coal or other carbonaceous material. Therefore, as the chemist calls carbon monoxide, incorrectly, carbonic oxide, simply to avoid confusion incident to the attempt to supersede a long standing error in terminology, we may, and almost must, adhere to the term water vapor.

A gas may be heated so that it will expend energy on account of the heating. This takes place if it is allowed to expand. Hence the heat required to raise the temperature of a gas free to expand involves two offices to be performed. A substance, which is the gas in question, is to be heated. This requires one portion of the heat. Then energy has to be supplied to the gas to enable it in turn to expend energy on its own expansion. This requires a second portion.

If the gas is confined so as to be incapable of expansion, the temperature can be more readily raised. The gas is inert and merely represents a mass to be heated. Less heat is required than in the case where the gas expands.

If it took a quantity of heat energy represented by 1 to heat a given weight of unexpanding gas a given amount, to heat the same weight of gas the same amount, when the gas is free to expand under its effect, would require a quantity of heat energy represented by 1·4058.

The quantity of heat required to heat identical weights of different solids, liquids or gases under identical conditions varies. The relative quantities required are termed the specific heats of the substances in question. The two kinds of specific heats of gases which have just been described are termed specific heat under constant volume and specific heat under constant pressure.

The same two kinds of specific heats exist for solids and liquids. The expansive force exerted by the latter when heated is so enormous that there is no practical way of accurately determining the specific heat at constant volume of most liquids or solids, because neither can be kept at a constant volume except in a very few instances.

Specific heat is, as has been said, the relative quantities of heat required to produce an identical change in temperature in equal quantities of different substances. The laws of specific heat vary in the cases of matter in the solid, liquid or gaseous state, and also vary with the temperature. In liquids and solids there is no approach to regularity. Water is taken as the standard, and the specific heat of liquids is stated by weight. Water has a very high specific heat. Mercury, for instance, has but approximately one-thirtieth the specific heat of water. A pound of water at a high temperature would have as much heating power in its cooling as would thirty pounds of mercury.

When we come to elements, we at once find a law

which is approximately followed. If we multiply the atomic weight of an elementary substance, such as gold, silver, lead, etc., by its specific heat, we get a number which is almost constant for all of the solid elements. This indicates that the heat required to heat an atom of a substance a given amount is approximately the same, of whatever element the atom may be.

The atomic weights of elements represent the relative weights of single atoms of the bodies in question or of equal numbers of atoms. It follows that if we take the ordinary specific heats, which are referred to equal weights of the substances, and multiply them by the atomic weights of the respective elements, the product will give the specific heat referred to, the heating of weights corresponding to the weights of the atoms.

These products are termed the atomic heats, and they vary but slightly among themselves. They are so nearly the same that a law was enunciated by Dulong and Petit to the effect that the atomic specific heats of the elements are identical.

Like many other enunciated laws, it does not hold true. The products given by the required multiplications vary from 5·39 to 6·87, and it is not easy to reconcile one's self to the idea that the differences are due to experimental error. The law is best accepted as being, like many other natural laws, only approximately true, and as being a useful instrument in determining certain chemical constants.

There are two expressions in constant use in thermodynamics which should be explained in this book, as they occur in discussions of the problems of expansion and contraction of gases. Once explained, the explanation may be easily remembered as being descriptions of near relatives of the two specific heats which have been described. The two specific heats were specific heat at constant volume and specific heat at constant pressure. The two expressions to be explained are adiabatic and isothermic expansion or contraction.

Suppose that a gas is placed in a condition which permits it to expand. The molecules repel each other, they beat back and forth constantly, striving to augment the length of the paths they move over, so if the conditions permit expansion, the gas expands. In expanding it will exert energy, and the energy has to be supplied from some source. If none is supplied from an external source, the gas will fall in temperature, the energy will be drawn from the inherent heat of the gas itself. Imagine the almost theoretic case when the gas expands thus absolutely at the expense of its own heat. No heat has been added to it, the expansion is adiabatic.

The condition rarely exists in practice, except in approximation, because as we work with gases under confinement, there is a surrounding vessel of more or less heat-conducting material, the gases pass through pipes and valves, and are in constant contact with objects at various temperatures. But one case exists in which a gas is compressed without the use of any restraining or impelling mechanism, without piston and cylinder, and where the expansion is so rapid and of such short duration that the adiabatic condition is almost exactly obtained. It occurs in the sound wave.

When a sound is made in a gas, waves start from the center of sound disturbance, and travel through space at the rate of about a thousand feet a second. In a second there may be anywhere from nine or ten waves up to twenty or more thousand such waves within the range of human audition. Each wave is composed, not of up and down motions, as in a wave on the sea, but of a forward implusc of the particles, followed by a springing back. On the forward impulse the air is compressed; on the reverse impulse, expanded. The action is very brief in duration and very slight, but the expansion and compression are practically adiabatic.

The air is surrounded by no containing vessel, and is condensed against its own inertia, so that every disturbing condition is absent, such as metal or glass to be heated, and the shortness of the period contributes to the perfection of action. The phenomena of the propagation of sound in air are used to deduce the factor 1·4058 (page 65). The determination is based upon the assumption that air in the sound wave expands adiabatically.

Now suppose that the gas expanded just as before except that we added heat to it, so that as it expanded it kept exactly the initial temperature. If it was air expanding in a cylinder, we might have a fire heating the cylinder. The air would absorb heat as it expanded without rising in temperature. Although the expression is not generally applied to such a case, the heat would be as truly latent heat as is the heat of liquefaction or of vaporization. We might start with 1 cubic foot of air at a temperature of 100° and end with 2 cubic feet of air at the same temperature. Our fire would do the work of preventing an adiabatic fall of temperature.

Such expansion is called isothermic expansion.

Opposed to expansion is contraction. There is an adiabatic contraction in which a gas yielding to external energy diminishes in volume without imparting to anything the heat given it by the energy expended on it. It grows hotter. If the action is theoretically perfect, if it gives off absolutely none of the heat energy into which the mechanical energy exerted upon it has been converted, the contraction is adiabatic.

But the vessel in which the air is compressed may be cooled artificially, so as to keep the air at precisely the same temperature. A stream of water may circulate through a water jacket surrounding the vessel. The water may be assumed to absorb the heat. The contraction is isothermic if the cooling is so complete that no rise in temperature takes place.

The primitive idea of a steam engine, if we except Hero's reaction engine, is represented by a piston and cylinder. A little water is placed in the cylinder and under the piston and is boiled. The steam forces the piston upward. At the end of the stroke, the steam is cooled and condensed to water and the piston descends.

Now, to avoid complication, imagine the steam replaced by air. The air is heated. It expands, and heat is constantly applied till the stroke is partly completed, so as to keep the air at the same temperature. This much of its expansion is isothermic. Next it is left to itself, and without receiving any more heat, expands until the end of the stroke is reached. This is adiabatic expansion. It now returns or performs the return stroke, the first portion by isothermic contraction, the air being cooled and kept cool, and completes the return stroke by its own contraction, with ensuing rise of temperature, or by adiabatic contraction. We will assume it to return to exactly the temperature it started at before it was heated at all.

The course of operations started with the air at a given volume and temperature, it went through a cycle of changes, and returned to its original volume and temperature, thus completing the cycle. To carry it out, conditions impossible of realization would have to be obtained. No engine could be built which would give the cycle perfectly.

An engine operating thus by expansion and contraction of a gas is a reversible engine. The steam engine is a typical example. The gas engine is another.

The cycle is termed Carnot's cycle, and the supposititious engine that would carry it out is called Carnot's engine. Such a cycle represents the most economical conditions under which power can be generated by heat. But the engine will never be built.

By following out the theory of Carnot's cycle, we reach the following law, the famous second law of thermo-dynamics:

In a reversible heat engine, the efficiency is represented by a fraction whose numerator is the range of temperature included in the operation of the engine, and whose denominator is the highest temperature included therein. These temperatures must be expressed in the absolute scale of temperature.

The law is all-important; directly or indirectly, it crops up constantly in the mechanics of the liquefaction of gases and of heat.

It has been stated thus:

Heat cannot of itself pass from a colder body to a hotter one, nor can it be made so to pass by any inanimate material mechanism, and no mechanism can be driven by a simple cooling of any material object below the temperature of surrounding objects. (Daniell.)

Another way of putting it is:

If the absolute temperature of a uniformly hot substance be divided into any number of equal parts, the effect of each of those parts in causing work to be performed is equal. (Rankine.)

If we indicate absolute temperature by Θ, and let Θ^1 and Θ^2 indicate two temperatures, Θ^1 being the higher, the second law states that in a reversible heat engine—

$$\text{Efficiency} = \frac{\Theta^1 - \Theta^2}{\Theta^1}$$

Mechanical energy can be expended and can develop heat energy, but heat energy can never develop in the mechanical form but a portion of its own quantity of energy. More and more mechanical energy is being converted into heat energy, and only a small portion can ever be recovered. Everything in the world tends to get to the same temperature; equalization of temperature is constantly taking place. In the existence of coal and air we have a form of potential energy, a potential high temperature. But even this potential high temperature is disappearing as coal is burned up. The available energy of the world gets less and less. The total energy is invariable.

The second law of thermo-dynamics leads us to the same conclusions as does the doctrine of the conservation of energy, although in this lowering of the scale of the world's energies, and the rendering them unavailable by man, there seems to be involved a contradiction of conservation of energy. But energy is intact in amount; in lowering its pitch, as we may express it, it ceases to be utilizable by man.

Liquid air, once produced, enables us to utilize heat which otherwise would be unavailable. The trouble is that to produce liquid air we have hitherto been obliged to expend a great deal more available energy than we can utilize of normally unavailable energy by its gasification.

Matter, as it exists in three states, solid, liquid and gaseous, is subject to two changes of state. Melting is one of these changes, when it changes from the solid to the liquid state; vaporization is another, when it changes from the liquid to the gaseous state.

Energy has to be used to bring about such changes of state, and no insignificant amounts, but very large amounts, relatively speaking, must be expended to effect the changes. Such energy is usually applied in the form of heat. If we wish to apply energy to a lump of ice and change it to the liquid state, we place it in a vessel upon a hot stove. If we wish to apply energy to the water so produced and change it to the gaseous state, we keep it on the stove, and presently it boils.

By measuring the heat applied, it is found that a great deal is required to change the solid into a liquid and the liquid into a gas. This is not all.

If we put a lump of ice into water, the water always takes the same temperature and keeps it until the last bit of ice is melted, provided that time is given for the water to assume the given temperature. We may apply heat to the water. If it were plain water, or if it were water with some unliquefiable solid floating in it, such as a lump of cork or a block of wood, every addition of heat would show itself in a rise of the thermometer. But as long as the ice is floating about in it the water will be practically unchanged in temperature, and will come back to the original temperature from any slight departure therefrom, as soon as taken from the fire, so that the ice has time to act upon it. Suppose we have put a pound of ice into the vessel. To melt it will require as much heat as would raise a pound of water nearly to the boiling point.

Imagine a pound of ice just ready to melt put into one vessel and a pound of water into another. If both were equally hot, their temperature would be 0° C. (32° F.) Now imagine exactly the same amount of heat applied to both until the ice was completely melted. We started with a pound of ice at 0° C. We should find at the end of the process that we had a pound of water at exactly the same temperature in the place of the ice. Meanwhile what would have happened to the water in the other vessel? It would have become so hot that the hand could not endure the heat. It would have taken a temperature of 80° C. (176° F.)

We have seen that our forefathers were not so fond of the term energy as we are. The ideas of the scientific world were not so well formulated as now, and the inevitable result followed that there was more complexity grafted upon the natural order of things than was necessary. They found that a quantity of heat was required to melt ice, and that it melted it without raising the temperature. The temperature would only begin to rise after the ice was melted. So they said the heat lies hidden; as it did not show itself on the thermometer scale, it must be concealed from us. They called it *Latent Heat*, which means hidden heat.

A similar, but more pronounced, disappearance of heat takes place when water is made into gas, when we boil it in a kettle or boiler. The heat required to convert a pound of water into steam at atmospheric pressure would raise the temperature of ten pounds of water 54° C. (97·2° F.) Suppose that the water we proposed to boil off had the temperature of 46° C. (115° F.) when we started. This would be a heat which the hand could comfortably bear. Then it is obvious that after enough heat had been applied it would reach the temperature of 100° C. (212° F.) A thing heated is supposed to grow hotter, and our water would act as it ought to do. But once the temperature of 100° C. (212° F.) was reached, the water would no longer grow hot. It would stay at the temperature named, it would begin to boil, and would gradually grow less and less in volume, and without the heat increasing, each particle would require ten times the heat expended on its preliminary heating to be converted into steam or gaseous water. The temperature of the steam, however, would be 100° C. (212° F.)

These are examples of the two most prominent latent heats, the latent heat of fusion and of vaporization. The term is so convenient that it will be used for a long time to come. The better term would be the energy of melting or of fusion and the energy of vaporization.

When a gas expands, it practically always expends energy and grows cold. Therefore, in the expansion of a gas und r ordinary conditions, a loss of heat occurs, so that a third kind of latent heat may be assumed to exist, the latent heat of expansion against pressure. This, however, is an expression not much used, and it is in the relation of specific heats at constant volume and at constant pressure that the conception finds its nearest expression.

We use ice to cool our drinking water, and perhaps never give a thought to the phenomena manifested. Yet it is very impressive to see how a small lump of ice can cool a large pitcher of water. In melting, it can reduce four times its own weight of water from the temperature of a living room to that of freezing, and as long as a particle of ice is left, the water will remain cold. A lump of ice, weighing one hundred pounds, lasts for a long time in a refrigerator. It absorbs as much heat in melting as would heat a ton of water through several degrees of the thermometric scale.

If circumstances are such as to produce vaporization at ordinary temperatures, the substance vaporized must absorb heat energy. A cloth wet with alcohol dries rapidly, because alcohol vaporizes or is converted into gas at ordinary temperatures. Heat is absorbed, and the cloth becomes very cold. In the tropics drinking water is kept in porous vessels. It exudes to the surface and evaporates therefrom. Heat is absorbed in the process, and the water gets cool. A workman employed in steel works cannot endure the heat of the furnaces and metal until he perspires heavily, and then he is comfortable. Irrespective of the physiological aspect of the case, the heavy perspiration by the heat energy absorbed in its evaporation keeps the skin from scorching. If he ceases from any cause to perspire profusely, he has to stop work until the sudorific glands begin to work once more.

Evaporation, which is slow boiling, here effects a cooling of the water and of the perspiring workman.

The term "boiling" is so firmly rooted in the mind as an expression of heat that it is a little hard to think of it as indicating cold. Repeatedly we read of experimenters with liquefied gases using a vacuum so as to make a gas boil and thereby produce cold. One might think that anything which would make a gas boil would be heat.

If what has been said about latent heat has been read and understood, it will be seen that boiling is a cooling process. If we wet the finger and hold it in a draught it becomes cold, because the water evaporates or boils off. This is a practical proof. But if it were possible to heat water so that it would not boil, the temperature of a pound of water would rise close to a red heat if enough heat were applied to it to boil it away under ordinary conditions. In other words, boiling keeps water relatively cool; it cannot get hotter under atmospheric pressure than 100° C. (212° F.)

The way in which water is made to boil is usually by applying heat to it. A very familiar old experiment may be cited where cold is applied, producing a vacuum, and the simple vacuum causes strong ebullition. A round-bottom flask is half filled with water, and it is brought to the boil and kept so until the upper half of the flask is full of steam. It is removed from the source of heat, allowed to come to rest, and is then tightly corked and inverted. Cold water is poured over it. This condenses the steam, and forms a partial vacuum. The water which was quiescent now boils with great energy, because of the reduction of pressure, and its own temperature falls. If a thermometer had its bulb immersed in the water, a reduction of temperature would be indicated.

It is obvious that this application of a vacuum is a means of lowering temperature. It lowers it by causing water to boil, so that we find the boiling of water a synonym for cooling or reduction of temperature.

Substitute liquid ethylene, liquid air, or other liquefied gas for water and apply a vacuum. The liquid will boil with increased energy and vigor, and its temperature will fall. A boiling gas is a cooled gas and is used as a cooling or refrigerating agent.

No one ever thinks of boiling a gas by imparting artificial heat to it. It is done either by exposing it to the atmosphere or by exhausting the vessel in which it is contained. The exhaustion makes it boil harder than ever. Exposure to the temperature of a boiling gas is exposure to cold. The more intense the boiling is, the greater is the cold. This expresses the condition of things obtaining in the work we are to describe.

If we speak of a thing being exposed to the temperature of boiling oxygen, at atmospheric pressure, it is very cold; if to the temperature of oxygen boiling under exhaustion, it is still colder. If we speak of a gas being made to boil, it means that we apply exhaustion, and that its boiling is a synonym for its growing colder. The student of this subject must therefore associate boiling with coldness, and get rid of its old association with heat. He must realize that boiling is a cooling operation, that if it did not boil, the water in a tea-kettle would get several times hotter than it can in fact.

The spheroidal state of matter forms so important a subject, in connection with the liquefaction of gases, that it should be well understood by the reader. It is to our vision a very peculiar condition into which liquids sometimes enter. In reality it is their normal condition, and the reason it seems to us peculiar is because the conditions for breaking it up are so very generally present.

In a liquid there is a slight force of attraction between the molecules. Hence the interior molecules are drawn to one another and are subjected to equal pulling stresses in all directions. On the outside or on the surface of a liquid, the molecules are pulled right and left and inward. Hence the outside is in a state of strain and constantly wants to become of as small area as possible. By an elementary proposition of geometry we can prove that of all solids of equal volume, the sphere has the smallest superficial area. Hence, if a mass of liquid is perfectly free from all external influences, the outer surface, under the effects of the lateral pulling that goes on among the molecules, will shrink to the smallest possible area by drawing the liquid into the shape of a sphere.

A liquid so situated that it is drawn by its own surface film into a shape approximating a sphere is said to be in the spheroidal state. The surface film composed of molecules acts exactly like a thin membrane of india rubber.

When a liquid touches no solid or liquid, it takes the spheroidal shape. The free portion of a drop of water, dependent from a rod, is drawn by its enveloping film into a spheroidal shape. If another rod touches it, the spheroidal shape where they meet is destroyed. When a solid is wet by a liquid, it is because the molecules of the liquid have a greater attraction for the solid than they have for themselves. Hence the skin-like action of the outer layer of molecules is destroyed when a solid which the fluid can wet is brought into contact with them

Mercury wets very few substances. When thrown upon a non-metallic surface, or upon a metallic surface of iron or of some metal which it cannot wet, it forms, as it is scattered about, a quantity of minute globules. Each seems to be a minute ball rolling about freely. Yet they are perfectly liquid. The surface tension, or the elastic pulling of their surface layer of molecules, draws them into an approximately spherical form. If mercury is dropped upon silver, the spheroidal tendency is no longer discernible, because it makes a true contact with the silver, which destroys the spheroidal state.

If a liquid is placed upon a surface very much hotter than itself, it slowly evaporates, and the products of its evaporation form a sort of cushion upon which it lies out of contact with the hot substance. The formation of this cushion of vapor or gas is interesting. It forms what is known as a Crookes layer. It is named from Prof. William Crookes, of England, who discovered the characteristic phenomena of gases at high rarefactions.

When gases exist in the condition in question, which condition is sometimes called the radiant state, they are in so rarefied a state that their molecules, in their vibrations, rarely collide. A billiard ball pursues normally a straight course from cushion to cushion, unless it collides with another ball. This is what the molecules of a gas do. They keep a straight path until deflected from it by collision with other molecules. If a silver dish is heated quite hot, and a drop of water is placed in it, the drop becomes warm and evolves steam. The molecules of steam from its under surface, under the influence of the hot vessel, become hot and beat back and forth from drop to vessel. This distance is so small, and the paths of vibration of the molecules are so long on account of the heating, that very few collisions occur. The molecules simply repeat their paths up and down from drop to dish, and thus form a cushion which prevents the water from touching the dish. The water is drawn into an approximately spherical shape, and the spheroidal state appears.

The cushion formed by the non-colliding molecules is termed a Crookes layer. Because the molecules do not collide there is no tendency to drive the steam out laterally. There is probably a very small proportion which escapes at the sides. The diagram gives the ideal section of a drop of water resting on a Crookes layer. The real layer is exceedingly thin. The distance between water and vessel may be termed infinitesimally small.

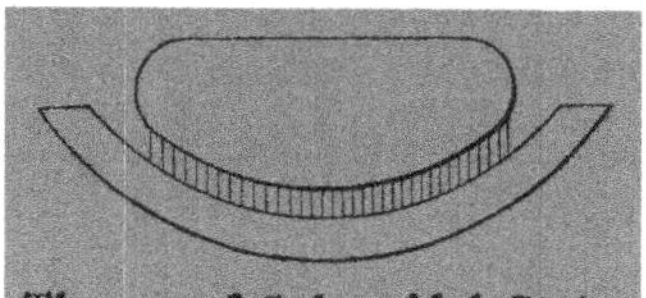
Theory of Spheroidal State.

A very homely simile would be afforded by a moving crowd. A man might elbow his way through it, and thereby thrust people to the right and left. But if the crowd was sparse enough, he would go right through it without pushing anyone laterally. In the Crookes layer the crowd of molecules is so sparse that the molecules do not hit and elbow each other. Therefore, there can be no side pressure, and the cushion of steam, in the experiment cited, stays under and supports the water.

It might be said that ordinary steam would form a cushion or layer between the water and hot metal. But it would not, because the weight of the water would squeeze it out and the water would touch the hot metal and would boil violently. But it is obvious that in a Crookes layer, where the particles of molecules do not collide, there is no possibility of their being squeezed out sideways, as there can be no side push upon them.

The experiment, as usually shown at lectures, is thus performed: A thick metal cup, preferably of silver, although brass is almost as good, is heated nearly or quite to redness. Water is now poured into it. Instead of bursting into violent ebullition, it lies in a shape like a flattened sphere, moving about constantly, but not boiling. The cup is allowed to cool, the water keeping the spheroidal state, but losing heat slowly. After a while the cup gets so cool that the water can touch the metal, which is still hot. Violent boiling begins and gets more and more violent, with a curious *crescendo* effect, until the water is reduced considerably in amount, when perhaps the small residue resumes the spheroidal condition for a few seconds more.

An excellent cup for the experiment can be made by hollowing out a thickish disk of brass. A round-ended cylinder of wood may be placed vertically upon it, the brass resting over a hole somewhat smaller than itself, bored in a block of wood. A blow with a hammer on the wooden cylinder will cup the brass sufficiently to make it hold water. It may be heated over a candle or alcohol lamp, and the water may be poured in from a spoon. A silver coin makes a still better cup. A long wire handle with one end thrust into a cork and the other bent into a ring will answer to hold it.

The demonstration, to show that the drop does not touch the metal, is illustrated in the cut. A drop of water rests on a Crookes layer over a hot flat silver plate. It may be projected by a magic lantern on the screen or may be looked at directly. In either case it is seen that light passes between drop and plate.

The importance of the spheroidal state in relation to the liquefaction of gases cannot be overestimated. It alone has rendered possible the achievement of the extraordinary results of the last few years. Except for the spheroidal state, it would be a matter of the greatest difficulty to manipulate liquid gases, and the perils of liquid air would be beyond estimate.

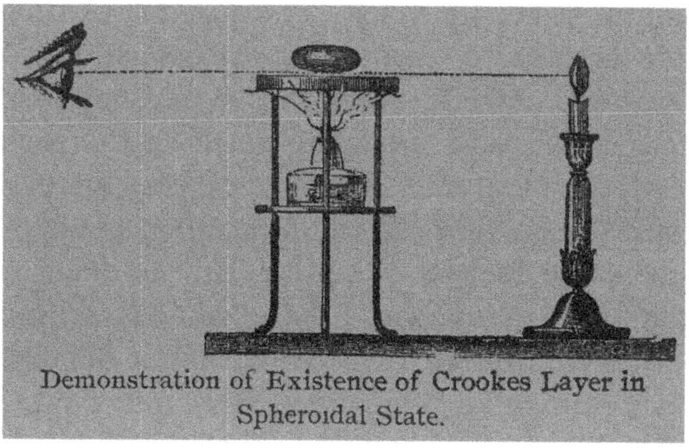

Demonstration of Existence of Crookes Layer in Spheroidal State.

But owing to the existence of the spheroidal state, and to its ready assumption by liquid gases, we are able to handle them much as we should water, although it is literally the same as if we kept water in red hot vessels. The experiments just described show how easy it is to do this. It is still easier to keep liquid air in vessels at atmospheric temperatures because the atmospheric temperature keeps our vessels, in a sense, almost red hot for liquid air. They are maintained at the temperature producing the spheroidal state without the need of any artificial source of heat.

A familiar experiment in the solidification of gases is the production of carbon dioxide snow. This intensely cold solid can be handled with impunity, it can be taken into the mouth, but does no harm, unless it is pressed against the skin, when it produces a bad blister from the intense cold. It is prevented from touching the skin by a Crookes layer, although it is hard to believe that a Crookes layer could support a solid of fixed shape on its cushion, but such must be the case. The support of the drop of water is easy to comprehend, because the drop flattens down until it is of the same shape as the body it rests on, and, adapting itself to the shape, is practically at even distance from it as concerns its lower surface, so that all the molecules have practically the same length of path. But to imagine an irregular lump of carbon dioxide snow so supported is not so easy, although we know that it occurs.

Yet a common experience is that many intensely cold objects can be handled without hurting the skin, and in many cases it is due to the spheroidal state, or at least to the formation of a Crookes layer.

CHAPTER IV.

PHYSICS AND CHEMISTRY OF AIR.

The atmosphere as an ocean—What air is—Its constituents—Relations of air to living beings—The chemist's and physicist's view of air—Its constancy of composition—Carbon dioxide—Oxygen—Nitrogen, argon and other constituents.

The physics of the atmosphere is very simple. The members of the animal world are often said to walk about on the bottom of an ocean of air, like crustaceans in the ocean of water. As fish swim about in the water of the actual ocean, so may birds and flying insects be noted as tenants of the atmosphere itself. There are, however, very great and fundamental differences; the analogy is a very incomplete one.

The fish and crustaceans live surrounded by a medium whose specific gravity is not far different from their own. A fish not only swims in water, but floats in it. By muscular contraction of his air bladder, he can increase his specific gravity so as to sink toward the bottom, or he can increase its size and rise toward the surface. Neither bird nor insect floats in equilibrium in the air. They are sustained by mechanical energy, derived partly from their own muscular system and partly, perhaps, by the internal energy of the air, due to variations in velocity of air currents. A crab has but the slightest hold upon the bottom of the water over which he crawls. Almost all his weight is buoyed up by the water. When he crawls on the shore, his legs have probably over eight hundred times as much weight in the concrete to deal with as when he is in the water.

Thus, our atmosphere has a far different relation to us than is held by the true ocean of liquid matter that spreads over so large a proportion of the earth's surface to its tenants. Its chemical constitution also is fundamentally different.

Water is a chemical compound, containing in chemical combination two elements, oxygen and hydrogen. The composition of its molecule is expressed by saying that it contains two atoms of hydrogen and one of oxygen. If water is decomposed, it resolves itself into two volumes of hydrogen to one volume of oxygen. A cubic inch of water will give about one and a half cubic feet of the gases named.

The atmosphere, the survivor of countless geologic ages, left after terrestrial changes of every kind, which has been warmed by centuries of sunlight, and which has been the theater of electric disturbances of the most violent kind, and which has been acted on by the tremendous vegetation of the carboniferous era, remains a simple mixture of gases, as far as its essential constituents are concerned. The constituents are not chemically combined, but are as free from any alliance with each other as the clay of the Mississippi and Missouri is from any fixed combination with the water that carries it in suspension toward the Gulf of Mexico.

For many years the composition of air has been given in text books as approximately consisting of one volume of oxygen and four volumes of nitrogen. This has proved an error. A chance discovery that nitrogen prepared from chemical sources had a different specific gravity from that prepared from the atmosphere was brilliantly utilized by the discoverers, Lord Rayleigh and Prof. Ramsay. They were engaged in physical research, and having lighted upon this very extraordinary fact, explained it by the discovery that a third element, argon, exists in air. It was a contribution from physics to chemistry. A chemist would not have had the audacity from purely chemical considerations to believe or suggest that an undiscovered element lay hidden in our atmosphere, and that we had breathed an unidentified gas, and had analyzed our air without finding it or suspecting its existence. The discovery was so revolutionary that it formed another step on the road to scientific credulity which we are traveling. Science has done so much that we are prepared to believe anything which may be attributed to her. Since 1894 other elements have been found in the air, and we find all our text books further invalidated in their descriptions of the very air we breathe.

Air is not a chemical combination, because its constituents have so little affinity for each other, and nitrogen has long been cited as an element of generally feeble affinities, and rather of the inert type. But it has to yield the palm to argon in this regard. The latter seems to be able to combine with nothing whatever.

Physiologically, our active relations with the air

concern only its oxygen, leaving aside impurities. We use the oxygen in our bodies to maintain life. The human system burns up the food it eats, and exerts energy of various kinds. The nitrogen and other elements act as diluents only. The animal system can do nothing with either of them.

An infinitesimal amount of nitrogen in chemical combination may have very grave effects. A fraction of a grain of strychnine, which has as an essential constituent a very small fraction of a grain of nitrogen, will kill a man. Without the nitrogen it would no longer be strychnine, and would be innocuous; so that in the case of this poisonous alkaloid, we find a small fraction of a grain of nitrogen an essential in a deadly composition.

Yet, in the case of the air, because of its nitrogen being in the free state, we breathe in and out of our lungs tons and tons of nitrogen, and it has no effect upon us whatever. It is only a diluent of the oxygen which we live upon.

A cubic foot of air weighs about 536 grains. It is generally taken as the basis of specific gravity of gases, which is a misfortune, because it is only a mixture, and has nothing essentially fixed in its composition. Yet it is rather remarkable that air always contains exactly the same proportions of its important constituents, and, therefore, always has the same specific gravity. There is nothing comparable to it in nature, if we regard it as what it essentially is—a fortuitous yet absolutely uniform and identical mixture of independent and uncombined gases.

The physicist can speak of air differently from

the chemist. For the first named it is an almost perfect gas, and he can speak of it as a typical gas. The chemist cannot do this. To him it is a mixture of gases, and he cannot term air a gas.

Air supports combustion and life, on account of the oxygen which it contains. If the quantity of the oxygen in a volume of air is increased, it will support combustion with much more vigor than in the ordinary state. This increase may be effected by adding oxygen or removing nitrogen, or mechanical pressure may do it. In either case combustion becomes more intense. In constructing foundations under water or under the water-level in soil, the engineer uses an inverted case, like a gigantic box. From it water is excluded by air pumped in it at high pressure, which may rise to fifty pounds pressure to the square inch. These structures are termed caissons, and in them where air is used, compressed up to three atmospheres excess of pressure, there is in one foot of the compressed air four times as much oxygen as under ordinary conditions. A piece of lighted paper, when blown out in such an atmosphere, will relight instantly. This mode of increasing the oxygen increases, also, the nitrogen. The combustion is not nearly as vivid as with artificially enriched air.

One would suppose that some difference in the composition of air would be possible under the conditions prevailing on the earth. It is being constantly drawn upon by animal life. Animals, in breathing it, rob it of a portion of its oxygen, and add carbon dioxide gas to it; the plant world adds to its oxygen and removes its carbon dioxide. Yet so constant are the mixing and disturbance to which it is exposed, that it proves the same when subjected to analysis, no matter where collected—practically the same, for there are slight variations which can be detected in the percentage of its impurities.

The principal one of these last named substances is carbonic acid gas or carbon dioxide gas. By the rules of chemical terminology, this gas should be called carbonic oxide, but a concession to long usage is made in its case, and the older names are adhered to. It is a product of animal respiration, and is a chemical compound, each molecule containing one atom of carbon and two of oxygen. It is about fifty per cent. heavier than air, but, by the law of diffusion, tends to mix itself with perfect evenness with the lighter air. It is a product of all combustion, our chimneys delivering quantities of it. An ocean steamer pours out from her funnels nearly a ton a minute. Dissolved in water, it gives it a slight flavor, and is an antidote to flatness of taste of the fluid. It makes soda water and aerated beverages in general sparkle and effervesce. It has played an important role in the liquefaction of gases. It has itself been one of the earliest ones experimented on with any degree of success, and has been liquefied on the comparatively large scale for many years. It has been a good object for experimenters to practice on in order to enable them to liquefy other gases which less readily succumb to pressure and cold.

Its history is not without its tragic side. There are many caves and wells in which it accumulates. To enter and remain in one of these means a speedy death by asphyxiation. Casks or vats in breweries get filled with it in the fermentation process, and many instances of death to workmen, who incautiously descended into them, are told of. In its liquefaction at least one fatal explosion has occurred, as we shall see later.

The liquid carbon dioxide possesses one very striking peculiarity. It cools so rapidly when released from confinement that it renders latent so much heat as to produce large quantities of carbon dioxide snow. Other liquids solidify in part when allowed to evaporate rapidly, but none does it with such facility as carbon dioxide.

When air is liquefied, a cloudy appearance is always presented, which is removed by filtering it through filter paper. This cloudiness is attributed to solid carbon dioxide disseminated like pulverized chalk through the liquid.

CHAPTER V.

THE ROYAL INSTITUTION OF ENGLAND.

The Royal Institution—Its origin and objects—Count Rumford—Sir Humphry Davy—The Pneumatic Institute—Davy's experiments in inhaling poisonous gases—His engagement as director of the Royal Institution—His views on the utility of liquefying gases.

The Royal Institution of England has been identified for more than three-quarters of a century with the liquefaction of gases. Davy, Faraday and Dewar have associated this line of research firmly with it. The recent investigations of Dewar and his associates have been performed in part in the laboratory where Faraday worked so patiently with his bent tubes and did work which appears of such extraordinary merit, when his limited appliances are considered.

The Royal Institution was founded in 1799. In 1796, Sir Thomas Bernard, the Rt. Rev. Shute Barrington, LL.D., William Wilberforce and Mr. Elliott founded the "Society for Bettering the Condition of the Poor." One of its principal objects was the establishment of an institution to teach the application of science to the advancement of the arts of life.

A select committee was appointed in 1799 to confer with Count Rumford on the matter, subscriptions were received, and the Royal Institution was established.

Count Rumford, who took such an interest in its

Courtesy of McClure's Magazine.

Laboratory of the Royal Institution.

organization, was an American, Benjamin Thompson by name, born in 1753, in Woburn, Mass. His life was a curious medley of diplomatic and army service and scientific study. He pretty thoroughly expatriated himself, his politics during the American revolution being on the Tory or Royalist side.

Courtesy of McClure's Magazine.

Lecture Room at the Royal Institution.

Yet Harvard College and the American Academy of Sciences were remembered in his will. He married the widow of Lavoisier, the famous French chemist, whose almost prophetic words on the lique faction of gases are proudly quoted by the French Academy of Sciences.

From an official copy of the charter and by-laws of the Royal Institution of Great Britain, dated 1835, we learn something of the early history of the foundation of the society.

It was legally established under a charter dated 1800, in the days of George the Third, and in 1810 its powers and functions were enlarged and confirmed by act of Parliament. It was a somewhat high priced society, as such things go. The entering member had to pay five guineas admission fee, and the annual dues were also five guineas. The entering member had to pay five guineas in addition to the above, to be devoted to the library or to some of the collections.

Mr. John Fuller was one of the great benefactors of the Institution. He established two professorships on foundations of £3,333 6s. 8d. each, which sums constitute two-thirds of £10,000, for which the Institution was his debtor.

The Fullerian Professorship of Chemistry is the one of most interest in connection with our subject. Its first incumbent was Michael Faraday. The chair was established in 1833, ten years after Faraday's first work on the liquefaction of gases. Faraday's appointment in the same year is chronicled in the pamphlet of 1835, just alluded to. The donor did not long survive his foundation of the chair. In the *Philosophical Magazine* for 1834, we find recorded a meeting of the Royal Institution, held on April 18 of that year, on account of the death of Mr. Fuller, who had done so much for the Institution.

Prof. James Dewar now occupies this chair.

Count Rumford had heard of the young scientist, Humphry Davy, and he engaged him a few years after the founding of the Institution, when only twenty-two years old, to be director. At first Count Rumford distrusted Davy and felt that he had been engaged precipitately. There were certain peculiarities about him which caused him to produce an unpleasant impression. But it very soon transpired that Davy was a most capable chemist, although it was impossible to foresee the renown he was destined to win for his country and for the Royal Institution. It is said that Count Rumford wished to find some one to give fame to the Institution. It soon appeared that he had made a most happy choice, and Davy gave it the most liberal meed of fame by his researches and discoveries.

Humphry Davy was born in Penzance, Cornwall, Eng., December 17, 1778. He early in life showed a great fondness for science. A Dr. Beddoes had established at Clifton, near Bristol, a sort of hospital for the investigation of the treatment of disease by the application of gases in general. It was entitled the Pneumatic Institution. Davy was engaged to be the superintendent and accepted, although he was but nineteen years old.

As we now look back upon Davy's early engagement, it is impossible to avoid feeling that the scheme in which he was embarked savored of strong peculiarity, to say no more. Yet he inspired it with rays from the lamp of true science, and thereby brought the genuineness of his character more strongly than ever to the front.

He was engaged to test the action of gases as remedial agents. He came very near proving their efficacy as a means of bringing about the death of subjects submitted to them. This was in his own person. He experimented by personally inhaling a number of different gases, a class of experiments which showed, in the state of science as it existed at that early day, the most intrepid courage. He experimented extensively with nitrous oxide or laughing gas. To test the combined effect of nitrous oxide and alcohol, he stupefied himself by drinking wine, and tried, as soon as he could collect himself, the effects of deep inhalations of nitrous oxide.

What he called nitrous gas was then tried, with rather disastrous results. We know now that, whether it was the lower or higher oxide, the ultimate effect of its reaction with the moisture of the mouth and mucous membrane would be to produce nitric acid within the system. This is exactly what his descriptions of the effects suggest. He burned his tongue and palate with it, it affected his teeth, and inflamed the mucous membrane. Then, not satisfied with this most disagreeable and dangerous experiment, he essayed what he called carbureted hydrogen.

This time he nearly died. He first, by expiration, got all the air possible out of his lungs and then inhaled what we know now to be a poison, or a mixture of poisons, as it probably contained carbon monoxide and carbon dioxide with hydrocarbons. The description of his sufferings and almost death is impressive, when read in the light of our present knowledge.

He tried carbon dioxide, but here Nature asserted

herself, and he could not get the pure gas into his lungs. Not to be beaten by the spasmodic closing of the epiglottis, he diluted the gas with air and tried it that way.

At twenty-two years of age we find him engaged by Count Rumford for the Royal Institution, indefatigably working in chemistry and physics, discovering the metals of the alkalis, producing the electric light, and after he had been but a few years in its service, doing one of the greatest services to science that ever fell to the lot of man to do—the engaging of Michael Faraday as his assistant in the Institution.

It is said that Davy's researches into the action of nitrous oxide or laughing gas on the human system were what led to his appointment to the Royal Institution.

Davy was very far-sighted in his views. He saw great possibilities in the liquefaction of gases. He said that it offered a way of impregnating water with gas without mechanical means. Soda water has since his time been made thus.

He said that great cold can be produced by liquid gases allowed to evaporate, and suggested the use of this faculty for preserving food. This outlines one of the cold storage processes, and it is hoped that liquid air may serve precisely the purpose outlined nearly eighty years ago by the great English philosopher.

Davy also had a great faith in the possibilities of liquefied gases as agents for generation of power. One of his papers (*Philosophical Transactions*, vol. xxiii., page 199) is devoted to this topic, and he gives figures to show what great power could be obtained from liquid carbon dioxide and the other gases which had been liquefied, and we find that, early in the life of the Royal Institution, Brunel tried the experiment of running an engine with liquefied carbon dioxide.

In connection with the subject of the liquefaction of gases, three names bring the Royal Institution prominently into notice: Davy, Faraday and Dewar. The first did comparatively little, but his suggestions were striking and suggestive.

The Royal Institution has struggled along for about a century, its centennial is at hand as this book goes to press, and the fine work done by Dewar and his associates in liquefying gases fitly marks the closing years of its first century of existence. Faraday's connection with it did more than was due merely to his far-reaching researches in chemistry and physics. The Institution has never been richly endowed, and for twenty-six years Faraday is said to have kept it alive by his lectures. He kept its accounts, and noted every expenditure down to the last farthing. The Institution gave him a fixed income of £100, and eventually the Fullerian professorship, appointing him for life, with the privilege of giving no lectures. The salary was then placed at £100.

In the same year the Institution was in trouble, and a committee reported on salaries, advising that no reduction should be made in Faraday's salary, "£100 per annum, house, coals and candles," which can only be taken as a compliment to the young scientist.

M Faraday

CHAPTER VI.

MICHAEL FARADAY.

Michael Faraday—His early life—Early devotion to science—His introduction to Humphry Davy—Attendance at scientific lectures—Engagement at the Royal Institution—Injuries from explosions in the laboratory—European tour with Davy—Rivalry of scientific men—Davy and Faraday as rivals—The liquefaction of chlorine—Davy's share in the experiment—Davy's opposition to Faraday's election as fellow of the Royal Society—Dr. Paris and the liquefaction of chlorine—Faraday's descriptions of his liquefactions—Explosions—Northmore's priority published by Faraday—Notes on Faraday's liquefaction of various gases—Exhibition of Thilorier's apparatus—Later work in liquefying gases—Discovery of the magnetism of oxygen gas—His death—Bent tubes as used by Faraday—Experiments with use of bent tubes—The Davy-Faraday laboratory.

Michael Faraday was born on September 22, 1791, at Newington, Surrey, England. His family was poor, with no pretensions to being in any but a low social level as society is organized and differentiated in England. His mother, who lived until 1838, was very proud of her son and his honors, although quite insufficiently educated to at all enter into his life's work. She was an excellent and extremely neat housekeeper. Faraday's education comprised little more than the rudiments of reading, writing and arithmetic. In 1804 he went as an errand boy to a bookseller, George Ribeau. Part of his work was the delivery of newspapers. Each copy circulated among a number of readers, for Ribeau lent the papers instead of selling them, and Faraday had to circulate in succession from house to house with the same copies.

In 1805 he began his apprenticeship as bookbinder and stationer, and at once began reading everything scientific that came in his way. He made simple experiments in chemistry, built an electric machine and other apparatus, and began to attend scientific lectures. In 1812 he heard four lectures by Sir Humphry Davy, and the same year he took an engagement as a journeyman bookbinder. The position was very disagreeable to him.

Before he had completed his seven years apprenticeship he took the step which shaped his whole life. He wrote to Sir Humphry Davy, asking for a position and sending elaborate notes of Davy's lectures which he had taken. He received a reply which he termed "immediate, kind and favorable," and early in March, 1813, he was engaged as assistant in the laboratory of the Royal Institution.

The histories of the early years of great men's lives are often of interest, and few exceed in this regard those of Faraday. Books were not so plentiful then as now, and Faraday used the opportunities which his trade of bookbinder and stationer put in his way to read scientific works. A series of letters by him written to his great friend Abbott show the tendency of his thoughts to chemistry, and incidentally show how indefinite were the theories on which the chemistry of that time was based; but Faraday's observations are often far in advance of the age. He speaks of the odor given off by metals when rubbed. Exactly this subject of odors, a very mysterious one, too, has been the topic of recent investigation. He objects to the names muriate of sodium and chlorate of sodium for common salt, and says that it should be called chloride of sodium, and sodium chloride is its name to-day. Another tendency of his mind was toward electricity. He gives the account of his making batteries, on the now old fashioned "pile" system, placing disks of zinc and copper, one upon the other, with paper moistened with acid between the alternate pairs. With these he decomposed water and acids and tells the results in the letters which have been preserved.

These letters, many of them written when he was but twenty years old, are wonderful examples of his intellectual powers. Here was a bookbinder's apprentice, but twenty years old, self-educated, speculating on subjects which constituted the most recondite branches of science and speculating rightly. The instances given above are but a few out of many which could be cited to show the precocity of his genius.

He kept a note book in which he entered the names and abstracts of articles in books and journals which had interested him. Sir Humphry Davy appears in it, for in this note book is the entry:

"Galvanism.—Mr. Davy has announced to the Royal Society a great discovery in chemistry—the fixed alkalis have been decomposed by the galvanic battery." This he credits to the *Chemical Observer*.

The greatest achievement of Sir Humphry Davy's

long career is noted by the humble apprentice, who was destined to succeed the older master and to equal or exceed him in renown.

An interesting illustration of Faraday's thoroughness occurred when he was but nineteen years old. He had attended some lectures given by Mr. Tatum on natural philosophy. They were given at his residence, 53 Dorset Street, Fleet Street. To enable him to do justice to the illustration of these lectures he actually learned perspective, doing all the drawings in a quarto treatise on this subject.

In this early work we recognize a threefold bent of his mind, always discernible in his long life's work. Chemistry was the branch of science which first claimed his attention and electricity was the work which he took up later in life. Chemistry and electricity, it will be remembered, were the two principal studies of his youthful days. The third subject which interested him was lecturing, and early in life we find him a lecturer in the Royal Institution, and for year after year he lectured there, and held a higher position than perhaps has ever been awarded an English speaking scientific lecturer. He also wrote upon the subject of lecturing and on the methods which should be followed in addressing audiences. He comes to the same conclusion which has so often been reached since—that a popular lecture will not be a good scientific one and that the converse also holds. From passages often quite long which refer to lecturing, the conclusion is drawn that he gave a great deal of thought to the subject and desired to achieve success in it.

On March 1, 1813, Faraday was engaged as assistant at the Royal Institution, at the salary of 25 shillings a week and the use of two living rooms at the top of the building. At once he began his initiation into serious work by assisting Davy in investigations into the properties of chloride of nitrogen, one of the worst explosives known to man. He chronicles explosion after explosion with it, his hand is torn open, his eyelid is cut; Sir H., as he calls Davy, has his hand bruised. They try to distill it, and it explodes, and Davy gets the worst of it, his face being cut in several places. They know the danger they are in, and wear glass masks, and Faraday at last says that "It is, as I before said, improper to consider it at any time as secure."

The dangers of science are appropriate to our subject. The liquefaction and compression of gases have given rise to many explosions, and to one of the worst explosions that has ever happened to an experimenter. We shall see later how Faraday and others suffered in experiments in these fields.

On October 13 of the same year, Sir Humphry Davy started on a tour over the Continent, on which Faraday was to accompany him. At the last moment Davy's valet refused to go, and Faraday agreed to do certain things which more properly would have fallen to the lot of that functionary. This arrangement, it was understood, was only to last until Paris was reached. In reality Davy completed the tour without any valet, and Faraday shrewdly concluded that finally he preferred to do without one.

In the early days of science there was a much greater spirit of rivalry among scientific men than at the present time. Seventy or eighty years ago there was a comparatively small body of scientific facts in the possession of man. The initial steps toward the acquirement of this knowledge had been made, and the acquirement and recording of facts proceeded more and more rapidly every year, until at present we have been presented with amazing developments, one after another, which in their rapid succession have almost robbed us of the capability of being surprised.

In reading the quaint story of the life of the bookbinder's apprentice Faraday, and of his experiences with Sir Humphry Davy during their continental tour, it is easy to perceive a sort of overriding tendency on the part of the older philosopher whose assistant he was. Faraday, too, had something to complain of from Lady Davy, but he seems to have held his own with her. Faraday's complaint was that he was requested to do certain things on the tour which he had not undertaken to do and against doing which he protested. At intervals after this journey, which took place in 1813–14, while he was twenty-two and twenty-three years old, some notes of discord can be heard, and the culmination seems to have been definitely reached in 1823. We have little to do with the unpleasantness between Faraday and Sir Humphry Davy; so we may briefly dispose of it now.

In 1823 Faraday did his first work on the liquefaction of gases. He liquefied chlorine and published the result, eventually disclaiming priority in favor of another investigator, Northmore, whose work is recorded later in this book.

On May 1, 1823, he was proposed for a fellow of

the Royal Society, of which Sir Humphry Davy was president. Faraday had succeeded by following Davy's suggestions in liquefying chlorine. Davy had not told him that liquefaction of chlorine was to be anticipated in carrying out his suggestions, and it was liquefied and identified as chlorine in Davy's absence. The work was therefore Faraday's own. Yet Sir Humphry Davy seems to have been jealous that part of the credit should attach to his junior associate. At any rate, it is definitely certain that Davy opposed Faraday's election as a fellow of the Royal Society, and actually asked him to withdraw the paper of nomination. Faraday said that, as the paper had been posted by his proposers, he could not take it down, and, on a further request, said that he knew that his proposers would not take it down. Then Davy said that he, as president, would take it down.

One of Faraday's proposers afterward told him that Davy spent an hour arguing that Faraday should not be elected. The certificate of his proposers had to be read at ten meetings. On the final ballot there was only one black ball. It is to be hoped that it was not thrown in by Sir Humphry Davy. After this Faraday and Davy got on more smoothly in all their relations. The culmination of their troubles seemed to mark the end of disturbance.

Thus Faraday's connection with the liquefaction of gases is concerned with one of the more important episodes of his life.

A gossipy life of Sir Humphry Davy has been written by Dr. John Ayrton Paris, who was an intimate friend of the philosopher, and who seems to have had a fancy for natural science. He was the first person to witness the liquefaction of chlorine by Faraday. The passage from his life of Davy in which he describes it is well worth transcribing, if only for the picture it gives us of the scientific life of those days. Dr. Paris had been invited to dinner with Sir Humphry Davy to meet the Rev. Uriah Tonkin. Sir Humphry had just set Faraday to work heating chlorine hydrate in a closed tube. We can see in our minds the brilliant company assembled at Sir Humphry's for dinner, while, not far away, Faraday, alone in the laboratory, was heating his chemical in a sealed tube, in imminent danger of blowing his eyes out. We can see Davy's biographer, dressed for dinner, standing by the side of the ex-bookbinder in his laboratory garb, watching and commenting on the operations of the master-hand. We can do no better than let Paris himself tell the story of Faraday's liquefaction of the gas chlorine:

"I had been invited to dine with Sir Humphry Davy on Wednesday, the 5th of March, 1823, for the purpose of meeting the Rev. Uriah Tonkin, the heir of his early friend and benefactor of that name. On quitting my house for that purpose, I perceived that I had time to spare, and I accordingly called on my way at the Royal Institution. Upon descending into the laboratory, I found Mr. Faraday engaged in experiments on chlorine and its hydrate in closed tubes. It appeared to me that the tube in which he was operating upon this substance contained some oily matter, and I rallied him upon the carelessness of employing soiled vessels. Mr. Faraday, upon inspecting the tube, acknowledged the justice of my remark, and expressed his surprise at the circumstance; in consequence of which he immediately proceeded to file off the sealed end, when, to our great astonishment, the contents suddenly exploded and the oily matter vanished.

"Mr. Faraday was completely at a loss to explain the occurrence, and proceeded to repeat the experiment with a view to its elucidation. I was unable, however, to remain and witness the result.

"Upon mentioning the circumstance to Sir Humphry Davy after dinner, he appeared much surprised; and, after a few moments of apparent abstraction, he said, 'I shall inquire about this experiment to-morrow.'

"Early on the next morning I received from Mr. Faraday the following laconic note:

"'Dear Sir: The *oil* you noticed yesterday turns out to be liquid chlorine.

"'Yours faithfully,
"'Michael Faraday.'"

It is seldom that we find such an interesting sidelight thrown upon the pages of early scientific history. It is a contribution to the everyday life of the old London world for which we cannot be too grateful to Dr. Paris. It reads like a bit out of Pepys' Diary. The unprejudiced reader of the present day will envy Dr. Paris his interview with Faraday, and few will feel that the meeting with the Rev. Uriah Tonkin should excite the same feeling to as great a degree.

In Faraday's letters we find several references to his work on the liquefaction of gases. In 1823 he had received from Davy the suggestion mentioned above to heat hydrate of chlorine in a sealed glass tube. This he did, and the fluid separated into two layers, and Faraday identified the lower layer as true liquid chlorine. He, to confirm this, compressed some chlorine gas in a tube, sealed it, cooled it, and again obtained liquid chlorine. The latter gas was dried before compression, so as to make the experiment absolutely conclusive.

He was troubled by his tubes bursting. His eyes were once burnt, another time were cut. He speaks of them as being filled with broken glass, the explosion being so violent as to drive pieces of glass through the window panes, "like pistol-shot," he writes.

This was in 1823. He found, on investigation, that neither he nor Sir Humphry Davy had priority in condensing gases into liquids, and so he published the article spoken of elsewhere (page 118) telling of Northmore's work.

In a letter written in 1836 he refers to Monge and Clouet's liquefaction of sulphur dioxide probably before 1800. This gas Faraday prepared by treating mercury with concentrated sulphuric acid, and found no difficulty in liquefying it. He attached credence to Monge and Clouet's very doubtful record, because he found the liquefaction of sulphurous oxide such an easy experiment to perform.

Sulphureted hydrogen he made in a sealed tube by first pouring into it some hydrochloric acid. Over this he placed a piece of platinum foil, and on this placed iron sulphide. The tube was then sealed, the acid was brought into contact with the sulphide of iron, and the tube was left for some days for the acid to act upon the sulphide. If necessary, the filled end of the tube was heated while the other end was cooled. He obtained a very limpid, clear fluid, whose specific gravity he puts at about 0·90.

When he came to experiment with carbon dioxide gas, he was badly troubled by explosions. He prepared it from ammonium carbonate and concentrated sulphuric acid. He credits it with requiring 36 atmospheres at 0° C. (32° F.) for liquefaction.

Euchlorine, as it was then called, he made by acting on potassium chlorate with sulphuric acid. After twenty-four hours' standing he heated the mixture to nearly 38° C. (100° F.), cooling the other end of the tube to —16° C. (3° F.) and condensing a dark yellow fluid.

Nitrous oxide or laughing gas was prepared by heating ammonium nitrate. This he heated first to partial decomposition, in order to get it as dry as possible. The procedure was rather superfluous, as in the decomposition water is inevitably produced, no matter how dry the salt is. Again he was troubled with explosions, for he got the pressure up to 50 atmospheres at 7·2° C. (45° F.)

Cyanogen he produced by heating dry mercury cyanide in one end of the sealed tube, and the cyanogen was condensed as a liquid in the other end.

Ammoniacal gas was absorbed by silver chloride. He found that 100 grains of silver chloride would absorb 130 cubic inches of the gas. This highly charged salt of silver, heated in the sealed tube, evolved ammonia in abundance, and he liquefied it without trouble.

Hydrochloric acid was made from ammonium chloride and sulphuric acid, and liquefied.

This was the work done in 1823. In another place will be found a full description of the bent tubes used by Faraday to liquefy gases. These tubes are still useful in demonstrations and for tests on the small scale, although their use is not free from danger.

It is reported by Prof. James Dewar, of the Royal Institution, that it appears from old papers or records that in 1838 Faraday exhibited at the Royal Institution Thilorier's apparatus for the liquefaction of carbon dioxide, lent him by Mr. Graham. This was a few years only after its first construction by the French scientist. The date of the lecture in which it was exhibited by Faraday was May 18, 1838. The exact date is recorded in the *Philosophical Magazine*, vol. xii., 1838, page 536.

With the exception of this incident, we have to pass over a long period, some twenty-odd years, before we find Faraday again seriously occupied with the liquefaction of gases. When past his fiftieth year he returned to the subject. He had then done much of his life's work, he had formulated theories of electricity, especially in relation to magnetism, and was in the midst of the electric studies of his life, which lasted until 1855. His work in electricity underlies all the amazing developments of the last two decades, and the action of the magnetic circuit and the production of definite voltages from dynamo-electric generators were never brought to an intelligible condition except by the use of lines of force, and these were a device of Faraday's, which enabled him to picture in his mind the action of a magnetic field of force upon a conductor swept through it.

To return to the liquefaction of gases, it was in 1845 that he began anew to try to liquefy various gases, and the results are embodied in a paper presented to the Royal Society and published in abstract in the Abstracts of the Papers communicated to the Royal Society of London under date of January 16, 1845. Some additional remarks on the same subject are given in the same volume under date of February 20.

He combined mechanical compression with cooling, using two air pumps, working in succession, one after the other, reminding us of Pictet's pumps, described on page 165. The first one had a cylinder one inch in diameter. The next pump, whose cylinder was one-half inch in diameter, took the compressed gas from the first one and gave a second compression. The gases were pumped into green bottle-glass tubes, one-sixth to one-quarter inch in external diameter. This seems a very small tube to employ, but the diameter is so stated in the abstract. The tubes were sealed at the upper end, which was, in some cases, bent downward so that it could be inserted into a cooling mixture. The pressure could be raised to fifty atmospheres. He sometimes used tubes closed with brass stopcocks.

The cold was produced by what he calls Thilorier's mixture of solid carbon dioxide and ether. This gave a temperature directly of —76·7° C. (—106° F.) To increase the cold he placed the mixture under an air pump and exhausted down to one twenty-sixth of an atmosphere. This gave him a temperature of —110° C. (—166° F.) His bath of carbon dioxide and ether, under these conditions, lasted only fifteen minutes.

He found that several gases condensed to liquids at the atmospheric temperature under this degree of refrigeration. Sometimes he preserved them by sealing up the tubes, and they remained liquid at ordinary temperatures. Others troubled him by their chemical action on the cement employed in connecting his apparatus. Some he succeeded in solidifying. These were sulphur dioxide, sulphureted hydrogen, nitrous oxide, hydriodic acid, hydrobromic acid and ammoniacal gas. He suggests the great availability of liquid nitrous oxide as a refrigerating agent.

It is interesting to note that he tried hydrogen and oxygen at 27 atmospheres, and failed to liquefy them. He also failed with nitrogen and nitric oxide at 50 atmospheres, carbon monoxide at 40 atmospheres, and coal gas at 32 atmospheres.

His work was greatly facilitated by the adoption of low temperatures. In his use of a volatile freezing mixture in a vacuum combined with mechanical pressure applied to the gas, we recognize the elements of the work of most of his successors in the work of liquefying gases.

Faraday did some of the greatest work of his life in the realm of electricity, and here we have to chronicle a discovery which is the basis of some very striking liquid air experiments. He found that not only iron and a few other metals are attracted by the magnet, but he found that the gas oxygen is highly magnetic, the discovery of this fact coming after Baucalari's discovery of the same. Baucalari was professor at Genoa. Faraday's date was 1847.

What strange exultation would have possessed his soul could he have seen liquid oxygen adhering in quantity to the pole-pieces of a magnet, and lying in a vessel over its poles and drawn by the attraction as if it were a veritable metal, although it is as far removed chemically from the metals as possible.

After a long life, one of the most touching and interesting in the history of science, Faraday felt his powers gradually failing. His life had few episodes outside of his scientific discoveries. Sir Humphry Davy, at last, did him justice; the disagreeable incident of 1823 was the last of its kind.

At the age of seventy-five, on August 25, 1867, he died. He had spent all his scientific life in the Royal Institution, and left it as a veritable legacy the story of his work on the liquefaction of gases, so ably prosecuted in the same building by Prof. Dewar.

Davy and Faraday are now commemorated by the Davy-Faraday Research Laboratory, in connection with the Royal Institution, founded by Dr. Ludwig Mond, which was opened in 1896.

CHAPTER VII.

EARLY EXPERIMENTERS AND THEIR METHODS.

Perkins' claim to have liquefied air—Its absurdity—Northmore's liquefaction of chlorine—Rumford's experiments as commented on by Faraday—Babbage's experiment in a drill hole in limestone rock—Monge and Clouet's alleged liquefaction of sulphurous oxide—Faraday's liquefaction of chlorine—Stromeyer's liquefaction of arseniureted hydrogen—Faraday's bent tubes for liquefaction of gases—Manometer for use with them—Experiment in a straight sealed tube in the liquefaction of chlorine—Davy's suggested method—Cagniard de la Tour—His bent tube experiments—D. Colladon—His apparatus as still preserved—Thilorier—His discovery of solid carbon dioxide—A fatal explosion—The improved Thilorier apparatus—Johann Natterer's apparatus—His experiments—Loir and Drion's solidification of carbon dioxide—Thomas Andrews, of Belfast.

The first hint of the liquefaction of air is given in the *Annals of Philosophy*, new series, vol. vi., page 66, 1823. It is merely a short note giving the title of a paper by Mr. Perkins. The paper was to be read at a meeting of the Royal Society, in 1823, but it was mislaid, and the Royal Society were spared the reading of it.

Mr. Perkins says that he exposed air to a pressure as high as 1,100 atmospheres, which is nearly eight tons to the square inch, or over half the pressure produced in a modern cannon. He says that the air upon compression disappeared, and left in its place a small quantity of liquid, permanent when the pressure was removed, tasteless, and without action on the skin. Faraday says (*Quarterly Journal*, xvi., page 240) "it resembled water," but thinks that it may be some unknown product of compressed air.

The present generation, to whom liquid air in quantity has become a plaything, recognize in Mr. Perkins' work a very simple state of things. The air disappeared because it all leaked out, and the water vapor present was condensed by the high pressure and was left in the apparatus. Had Faraday been given a sample of Perkins' "liquid air," he would at once have identified it as water.

In 1805 and 1806 papers by Thomas Northmore appeared in *Nicholson's Journal*, xii., page 368; xiii., page 233. Northmore was experimenting to see what effect pressure had upon a mixture of gases. He had a compression pump, mercury gauge and receivers, and pumped his gases directly into the receivers. He tried a metal receiver, but found it unsatisfactory and adopted a glass one.

Very fine illustrations of some of his screw connections, of his valve and of his siphon gauge, are given in the *Journal*. They show so little and such unimportant parts that it is rather surprising why such care was taken in so beautifully reproducing them.

He had all sorts of difficulties. His stopcocks troubled him, as they leaked. The metal parts of his pump corroded under the effect of the gases he experimented with, and his receivers exploded several times.

He condensed chlorine gas, then called oxygenated muriatic acid, describing the experiment as follows:

"Upon the compression of nearly two pints of oxygenated muriatic acid in a receiver two and a quarter cubic inches capacity, it speedily became converted into a yellow *fluid*."

He then comments upon its pungent odor and its great volatility.

He thinks that he liquefied sulphurous acid, but his pump piston became immovable very soon, on account of the action of the gas. He says that he obtained "a thick slimy fluid, of a dark yellow color." This, he claims, confirms Monge and Clouet's experiment, as given in Accum's "Chemistry," vol. i., page 319.

Faraday, whose mind was pre-eminently illumined and guided by the lamp of truth, contributed to the *Quarterly Journal of Science, Literature and the Arts*, vol. xvi., page 229 *et seq.*, a paper on the history of the condensation of gases. He states that when he liquefied chlorine gas a little earlier in the year 1823, he was unaware that "any of the class of bodies called *gases* had been reduced to the fluid form." He started an investigation into the history of the subject. He found that Count Rumford, in 1797, had exploded gunpowder in closed vessels and had claimed to confine the gases produced within the space previously occupied by the powder. This may, with all due respect to the distinguished inventor, be doubted Faraday speaks of the hissing sound observed when the products of combustion in Rumford's experiment were allowed to escape, and concludes that this may have been due to liquefied carbon dioxide.

The accepting a hissing sound as proof of liquefaction reminds us of Pictet's claim for the liquefaction and solidification of hydrogen, when so much was inferred from the noise due to the escaping of the stream and to its impinging on the floor.

Faraday does not make any point of the fact that carbon dioxide snow or solid carbon dioxide should have been produced. That this is formed when the liquid in question is permitted to evaporate under atmospheric pressure was unknown at the time the paper was written.

A most curious experiment on the decomposition of marble under pressure was made by Mr. Babbage in 1813. He wished to ascertain whether pressure would prevent chemical decomposition. The idea, in our days of high grade explosives, and when the recent explosions of liquid acetylene have done so much to bring a safe illuminant into evil repute, seems curious. But Mr. Babbage, with his inquiring mind, had a hole thirty inches deep and two inches wide drilled in the limestone rock at Chudley Rocks, Devonshire. A quantity of strong hydrochloric acid was poured into the hole, and a conical wooden plug, previously soaked in tallow, was driven into the mouth of the hole and the experimenters stood off and waited. They might be waiting yet, as far as the experiment went, for nothing occurred, the rock was not split and the plug was not expelled. Faraday thinks that liquid carbon dioxide may have been formed and lain quietly in the hole. He overlooks an important point—that the water of the hydrochloric acid would assist in lowering the pressure by its solvent action on the carbon dioxide. Mr. Babbage's conclusions are not given.

Faraday, in his paper on the history of the liquefaction of gases, says it is asserted that sulphurous acid gas had been liquefied by Monge and Clouet, but that he had not succeeded in finding any account of their process. Their work dates back to the end of the eighteenth century. Anyone who wishes to investigate the subject will find it clouded by uncertainty. On page 234 of the *Quarterly Journal of Science, Literature and the Arts*, vol. xvi., will be found references to seven authorities, and there seems to be no certainty obtainable from any of them. Faraday reaches the conclusion that the degrees of pressure and of cold required to liquefy sulphurous oxide are so slight that there is little doubt that Monge and Clouet did actually accomplish the experiment. The original authority cited for their work is Fourcroy, vol. ii., page 74. He states that the gas is liquefiable at "28° of cold." This temperature refers probably to the Centigrade scale, and reduces to —18·4° F.

The early experimenters had found that by exposing chlorine gas, produced by the usual methods, to cold, a solid substance was produced which was supposed to be solid chlorine. About 1810 this was examined by Sir Humphry Davy, who found it to be a compound of water and chlorine. Faraday analyzed it, and found it to contain approximately "27·7 chlorine, 72·3 water, or 1 proportional of chlorine and 10 of water."

Modern analysis but slightly changes Faraday's

figures, to chlorine 28 per cent., water 72 per cent., giving as a formula $Cl.OH_5$. The old investigators had not produced dry chlorine, and the substance which they cooled contained so much water that chlorine hydrate was produced by the refrigeration.

Sir Humphry Davy suggested that exposing the chlorine hydrate to heat under pressure would probably lead to some interesting results.

Without detailing Faraday's exact words, it may be enough to refer the reader to the *Philosophical Transactions of the Royal Society of London*, 1823, vol. xiii., page 160 *et seq.*—a most sumptuous publication wherein the work is described in full detail by Faraday. A subsequent note by Davy says that he thought one of three things might result from the experiment, and among them was the liquefaction of chlorine.

This was the origin of the last and most bitter dispute between Faraday and Davy. More is said of it on pages 106 and 107. After this the two lived on excellent terms. It must also be said that it was a very one-sided dispute, as far as any acrimony was concerned, Faraday showing not the least spirit of contention.

The assertion by Davy of what ideas were present in his mind when he suggested the experiment to his assistant was calculated to deprive Faraday of the entire glory of being the first to successfully liquefy chlorine. But the historical investigations of Faraday showed him that nearly twenty years earlier Northmore had made liquid chlorine, so that the bone of contention was pretty well disposed of.

Other less important liquefactions are that of

arseniureted hydrogen, claimed for Prof. Stromeyer, of Göttingen, in 1805, but very much doubted by Faraday (*Quarterly Journal*, xvi., page 236); and that of hydrochloric acid, claimed for Mr. Northmore, in 1805 (ibid., page 236; *Nicholson's Journal*, xii., page 368, xiii., page 232).

The arseniureted hydrogen experiment, however, has a great subjective interest, as it illustrates the danger inherent in the work of the early chemists. This gas is so frightfully poisonous that its discoverer is said to have been killed by inhaling a single bubble. Yet we read of Stromeyer producing it in quantity, by digesting an alloy of 15 parts tin and 1 of arsenic in strong muriatic acid, collecting it over the pueumatic trough, and exposing it to the temperature produced by mixing snow and calcium chloride, in which, as a test of its coldness, several pounds of quicksilver had been frozen in the course of a few minutes. This was certainly a most powerful freezing mixture. Yet Faraday doubts if the gas was really liquefied, as he himself had tried it at nearly —18° C. (0° F.) at a pressure of three atmospheres.

Had any accident happened during these experiments, had a retort burst or the high pressure apparatus exploded, the intrepid experimenters would have had a narrow escape with their lives, if they had not instantly succumbed to the poisonous gas.

As for hydrochloric acid gas, whose liquefaction had been claimed by Northmore in 1805, Faraday concludes that as 40 atmospheres pressure are required to liquefy it at an ordinary temperature, and as Northmore employed no cooling mixture, the supposed condensation did not take place.

For liquefying gases on the small scale when they can be evolved by heat, and at not too high pressures, the bent glass tubes devised by Faraday for this use may be employed. There are many shapes given by him, two of which are more directly in the line of our subject. One is applicable where no liquid is given off in the process of producing the gas, for it must be produced in the tube. Another is used where some liquid, such as water, is evolved during the gas evolution process.

The simple bent tube is shown in the cut. The tube as made is sealed at one end and bent in the middle. The gas-evolving material is placed in the closed end, and the other end, which has been left open for the introduction of the material, is closed after the introduction by melting the glass with a blow-pipe or Bunsen burner flame. In the construction of the tube care must be taken to maintain a good thickness of the glass where it is drawn out for closing. Often in drawing a tube down the glass becomes too thin for strength.

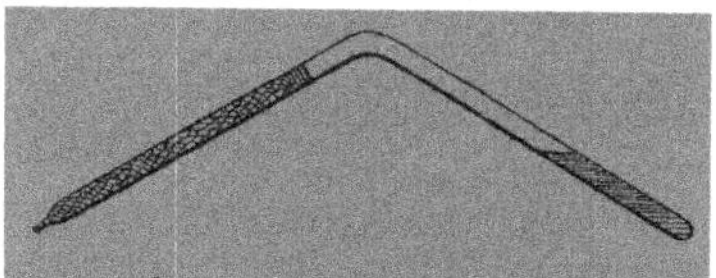
Faraday's Simple Bent Tube.

If the gas is one which liquefies by pressure alone, all that is necessary is to hold the tube in the position shown and heat the full end. As the gas is evolved it produces pressure in the tube, and if the pressure becomes great enough, and if the temperature of the empty end of the tube is cool enough, it liquefies and collects there in the liquid state.

But often cold is required in addition to pressure, and this is secured by inserting the empty end of the tube into a freezing mixture. Powdered ice and salt or powdered ice and calcium chloride are typical mixtures.

The other shape of tube is shown in the next cut. The tube was held inverted, as shown in the upper figure, and the substances were inserted, as shown, into one or both bends, *b* and *c*. A long-stemmed funnel was used to pour the liquids through, if liquids were used. The ends, *a* and *d*, were then sealed, and, by turning the tube over, everything collected in one end, *a*. The tube was placed with the empty end, *d*, in a freezing mixture. The end, *a*, was heated, if necessary. The liquefied gas collected in the further end, and any liquid that distilled over was caught in the intermediate bend.

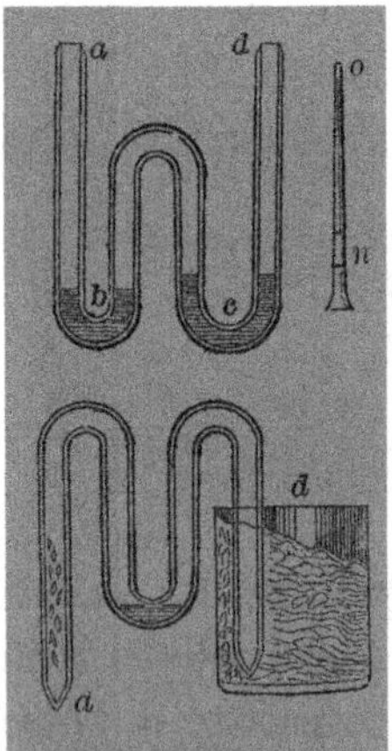

Faraday's Bent Tubes and Manometer.

To determine the pressure produced in the tube, a small tube closed at one end, *o*, and with a short bit of mercury, *n*, in its bore, was placed in the experimental tube before closing it. As the pressure rose, the mercury was forced toward the end of the small tube containing it. This it did because the air confined between the mercury and the top of the tube is compressed. If the distance from the mercury to the closed end of the tube is diminished to one-half its original length, and if the tube is of exactly even bore, it indicates a pressure of about fifteen pounds to the square inch in excess of the atmospheric pressure.

Faraday directs the manometers to be made of drawn-out tubing which is of greater diameter at the open than at the end which was to be closed. He directs that they be from eight to twelve inches long. They were calibrated and graduated by placing in them a drop of mercury. By careful manipulation this was moved from end to end of the tube and its length was marked off, step by step, for the whole length of the tube. This left the tube divided into lengths varying among themselves, but, as each corresponded to the volume of the same drop of mercury, each length would give an equal volume. By having the tube larger at the base than at the top the readings for high compression became more delicate. The mercury was left in the tube to act as an index ; the upper end of the tube was sealed after the graduation was ended.

In graduating the wide parts a larger quantity of mercury is prescribed for the operation, but the original divisions on the upper part of the tube gave the basis for its entire division.

In Faraday's "Chemical Manipulation," American edition, 1831, page 608, quite elaborate directions are given for making these gauges. It will be evident that considerable accuracy is attainable with them. By such a tube he states that it is easy to read off to above one hundred atmospheres.

In use he says that the compression tube should be bent in two places, giving three straight divisions, something like a letter N, and the little manometer is to be inserted in one of the divisions.

Seventy-five years ago Faraday, with such apparatus, liquefied chlorine, cyanogen, ammoniacal gas, carbonic acid gas and some others, as described in this and the two preceding chapters.

The greatest care is to be recommended in carrying out these experiments. The tubes are very prone to explode, and if they do, the explosion is very violent. A tube will sometimes be in part reduced to sand like grains of glass. There were many such explosions in the early days of chemistry, and the experimenters wore glass masks.

A very pretty experiment, which can be done in a straight closed tube, occurs in the liquefaction of chlorine from the hydrate. Chlorine hydrate, a compound of chlorine and water ($Cl.OH_5$), is made by saturating water with chlorine gas and surrounding the vessel containing it with ice. A somewhat strongly green-colored crystalline substance separates, which is chlorine hydrate. A more intense cold is needed to separate the crystals well.

A quantity of the crystals is placed in a tube closed at the bottom and the upper end is sealed. On heating the hydrate it melts. A purse-like drop of chlorine forms near the surface of the liquid and hangs therefrom down into the liquid, constantly increasing in size until it falls to the bottom and the fluid is divided into two layers. At the bottom is liquid chlorine, above it is water. By slight additional heat, if the other end of the tube is in a freezing mixture, chlorine can be distilled over, and will collect as a liquid in the cool end of the tube. The double bent tube may be used in this latter experiment.

Sir Humphry Davy, in 1823, suggested a modification of Faraday's process. He would fill the tube with the gas to be liquefied. Then a little water ether or alcohol is introduced, and the tube is sealed up. By heating the alcohol or other fluid, it gives off its vapor, and the pressure in the tube can thereby be brought up to any desired point within the limits fixed by the strength of the tube. In his own words, gas "is in one leg of a bent sealed tube, confined by mercury." The idea undoubtedly was to use such a tube as shown on page 124, and to place mercury in the intermediate bend, so as to shut off the water from the gas.

The idea is rather ingenious, but we cannot ascertain that it led to any results. Gas has practically always been compressed either by the pressure produced by its own evolution or by a pump or press, and Davy's suggestion has not been utilized to any extent.

Faraday thought so highly of the use of tubes in chemistry that a long chapter in his "Chemical Manipulation" is devoted to what he terms tube chemistry. It is illustrated with cuts representing many kinds of tubes, and the use of sealed tubes as here described for the liquefaction of gases forms only one of many applications which he describes.

To orientate ourselves we must note that the book in question appeared long before Faraday did his final work on the liquefaction of gases. In 1845 he used the two condensing pumps, one working into the other, for compressing gases, and condensed them in tubes made of green glass, and sometimes fitted with brass cocks at the ends. Thus he departed, in 1845, from the simplicity of manipulation which distinguished his work of 1823. The book whose American edition is dated 1831 is an exponent of his earlier and simpler methods.

Faraday has been used as a starting point in the history of the liquefaction of gases, because he not only is the first who did really thorough work on the subject, but because his investigations into the literature of the subject have greatly facilitated the fixing of the date of the work of the older investigators.

The earliest ideas about the possibility of liquefying gases were based very largely on the efficacy of pressure. The influence of pressure on liquefaction was not known, and various experimenters investigated it. M. le Baron Cagniard de la Tour attacked the subject, but in an inverted order. He demonstrated that liquids could be converted into gases of volume little more than twice their own. Had the scope of his work been properly appreciated, much trouble might have been spared more recent investigators. We know now that temperature is the essential thing in liquefying gas, and that pressure is altogether subsidiary to and dependent on it. There is no more impressive contrast to the work of the early investigators who devoted all their energies to the production of pressure for liquefying gases than the experiment described later (page 336), when the exterior surface of the simple tube of liquid air, exposed to exhaustion, drips with liquid air condensed from the atmosphere at atmospheric pressure by the intense cold.

It was just before Faraday liquefied chlorine that the baron did his work and established La Tour's law, that a liquid can be converted into a gas which shall not exceed in volume the liquid itself. At least, his investigations gave the proof of this fact so nearly that the law is thus stated under his name.

La Tour worked with sealed tubes, as did Faraday, partly filling tubes with various liquids and applying heat. He made a portion of the tube itself act as a manometer or pressure indicator.

Before beginning his more accurate work on the small scale with glass tubes, he tried an experiment on the large scale which reminds one of Otto Van Guericke's methods. The wonder is that he did not have an explosion.

His original papers were published in the *Annales de Chimie et de Physique* in 1822, and afford a good example of early methods of work.

He first took the end of a cannon and filled one-third of its interior volume with alcohol. In it he placed what he calls a ball of silex, and closed the gun hermetically. On shaking it, the ball was checked in its motion by the liquid. He applied heat gradually, and eventually reached a point when the ball bounced about without obstruction, as heard from the outside. Water was tried, but did not work so well. Petroleum naphtha (?) and ether acted like alcohol.

He unhesitatingly took it as proved that he had gasified alcohol, ether and petroleum naphtha in a space but three times their original volume. The fact that water did not give the same evidence operated in strong confirmation of his conclusions.

The next point was to obtain ocular evidence of the gasification of a liquid in such a limited space.

Accordingly he sealed up in glass tubes ether, alcohol and his petroleum naphtha, and, providing the tubes with long glass tails melted to them for handles, he heated them. As the heat rose the liquids expanded, sometimes to twice their original volume, and became very mobile, and suddenly disappeared as they became converted into gas.

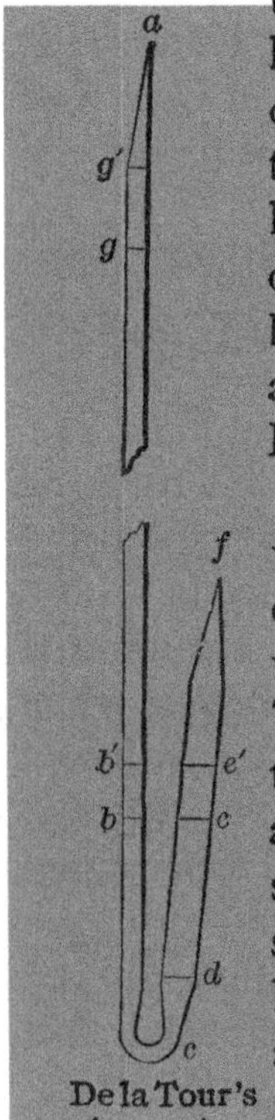

De la Tour's Apparatus.

The baron had this doubling of volume firmly fixed in his mind, for he filled his tubes about two-fifths full and succeeded in his experiment. Then he tried, with success, a tube nearly one-half filled, and another he filled a little over one-half with the liquid. This tube burst. He was careful also not to have any air mixed with the vapor of the liquids in his tubes.

His apparatus as shown in the cut was of the simplest description. Mercury was introduced before the tube was sealed. It settled in the bend, *c*. The liquid was poured into the large tube or bulb, and after expulsion of air, if desired, by boiling, the bulb was sealed. The other end, *a*, was also sealed. Now, the mercury lying in the bend had air above it in the small tube, and if the air changed in volume, the mercury, by rising or falling, would indicate the extent of such change. The liquids in the bulb were heated. As the pressure rose, the mercury was forced up the small tube, and the diminution of volume of the air gave the pressure. The air tube was so small in reference to the other that the mercury in rising made but little difference in the volume of the bulb.

The liquids, it will be observed, could not increase but two or three fold in volume. Any space in the right hand division of the tube not filled with the liquid contained its vapor. As the heat increased the liquid disappeared, being completely gasified, and eventually all the gas from one volume of liquid was contained in the space to the right of the mercury, in volume but two or three times the original volume.

The apparatus has various levels indicated in the original cut which the baron used in his description of his several experiments. Our illustration is a close reproduction of the original cut from the *Annales*, and, like Faraday's bent tubes, is an interesting example of early methods. The levels *c*, *e′*, *b*, *b′*, etc., indicate the levels assumed by the mercury as the conditions of volume of liquid, of gas, and the pressure in the tube varied.

He gives the details of several experiments. The details of a single one will be sufficient to give an idea of his methods of work.

In one experiment he filled the part of the tube marked *b*, *c*, *d*, *e*, in the cut on page 130 with mercury. The space above the mercury in the wide part of the tube was partly filled with ether. When all was in place, the ends, *a* and *f*, were sealed by melting the glass with a blowpipe. Heat was applied and the ether expanded and became gas, forcing the mercury up to the mark, *g*. The narrow portion of the tube contained the compressed air, and from its reduction in volume he calculated the pressure to which the gasified ether was subjected.

The tube of larger diameter was four and a quarter millimeters (about one-fifth of an inch), the smaller tube was one millimeter (about one twenty-fifth of an inch) in diameter. Care was taken to have this narrower tube of even diameter. Other marks of level, *e′*, *b′*, *g′*, are given by the baron to indicate how the experiment was performed with alcohol.

The experiments with different liquids are described, and some of his calculations show to advantage the work of a careful observer employing simple apparatus.

One tube was two-fifths filled with alcohol sp. gr. 0·844. The liquid expanded to double its volume and then, at a temperature of 258·7° C. (497·7° F.), suddenly disappeared. The pressure was about 119 atmospheres. Ether became gaseous at a temperature of 200° C. (392° F.) with a pressure of 37·5 atmospheres, the gas occupying twice the volume of the liquid.

Water became gaseous in four times its bulk at a temperature of about 412° C. (773·6° F.), or that of melting zinc. A little sodium carbonate had to be added to the water to prevent it from attacking the glass of the tube. Before he adopted this expedient his tubes broke when he used pure water in them.

Many years later we find Cailletet repeating this last experiment with pure water in a metallic tube.

As the vapors cooled, a cloudy appearance was observed, and the liquid, when the temperature fell sufficiently, suddenly reappeared.

La Tour was very near to obtaining the evidences

of the intermediate state observed by Thomas Andrews (page 147). As it was, he found that under the conditions a very wide departure from the law of relation of volumes of gases to their pressures existed, and he should be credited with a certain amount of important preparatory work in the liquefaction of gases, attacking the problem from the other end—effecting the gasification of liquids with very slight change of volume.

In Colladon's work, Geneva figures for the first time in the field we are treading. Later Pictet made the city by the lake famous by his liquefactions of gases.

Daniel Colladon, of Geneva, was the assistant of the great Ampère, and no apology is needed for inserting an incident in his life, as told by Raoul Pictet in his work "Etude Critique du Materialisme et du Spiritualisme." The last word is not to be rendered as our word "spiritualism;" in the French language it refers to the operations of the mind and soul, not to the fraudulent manifestations of so-called mediums.

Ampère had studied out his theory of magnetism, and had ordered apparatus to be made for its demonstration. A distinguished audience assembled for the lecture, and at the last moment the apparatus arrived from the maker.

Those who are familiar with the Ampère theory of magnetism know how it is demonstrated by wire bent into helices, which are poised like compass needles and are subject to the movements of a compass needle when an electric current is passed through them. When the current passes, the solenoids point north and south, the ends are attracted or expelled by one or the other pole of a magnet.

Ampère began his lecture and gave his demonstration on the blackboard. He was, in Prof. Pictet's words, "superb, eloquent in the power of his conviction." All present were delighted. Then the apparatus was taken in hand, and the practical proof of the theory was to be given.

The solenoids were mounted and connected to the electric terminals so that the current passed. *They refused to move.*

Ampère tried again, but in vain. The audience began to grow impatient. In the midst of the growing inquietude the suffering scientist did his best, but could get no result.

Ampère left the hall with Colladon. No one else was with them. They followed the Boulevard St. Michel toward the Seine, the tears running down Ampère's cheeks. He went to the house of an intimate friend, and tried to distract himself with a game of checkers—an old distraction with him.

Colladon now took the matter up, and began to reorganize the apparatus. He altered the method of suspension, substituting mercury cups for the solid contacts. He connected the electric terminals so that the currents passed, and all worked perfectly. It was eleven o'clock at night when he succeeded.

He ran to where Ampère was trying to forget his sorrows in checkers, and called out to the great scientist, gloomily studying his game, "It works, it goes, it moves!"

Ampère seized his hat, and the two rushed off to the laboratory, where it was so late that the porter wanted to exclude them from the laboratory. The scientist saw the experiments successfully performed as midnight crept over Paris.

The lecture was repeated to a wildly enthusiastic audience with the beautiful experimental demonstrations which have done so much to immortalize the name of Ampère.

As the audience left the hall, the Marquis de Laplace waited at the door until Colladon, the last to leave, was crossing the threshold. Laplace barred the way, extending his arms, and looked him in the face, and said:

"Young man, you did not give it the least little touch?"

Three or four years later, in 1828, Colladon, then corresponding member of the Academy of Science, performed many experiments in attempting to liquefy gases. His apparatus was almost exactly that of Cailletet, without the release cock, which was at the base of the success of the later experimenter.

The dimensions of the apparatus, in the metric system, are quoted on the cut, which is an exact reproduction of one given by Prof. Pictet in his article on his work in the liquefaction of gases.

Two shapes of the capillary tube are shown, for it was of importance to be able to introduce the end into a freezing mixture. The bending down of the end makes this more convenient than with a straight tube.

The tube within the steel reservoir was nearly an inch in external diameter. As it was exposed to the same pressure inside and outside, it could be made of thin glass. The thick walled capillary tube

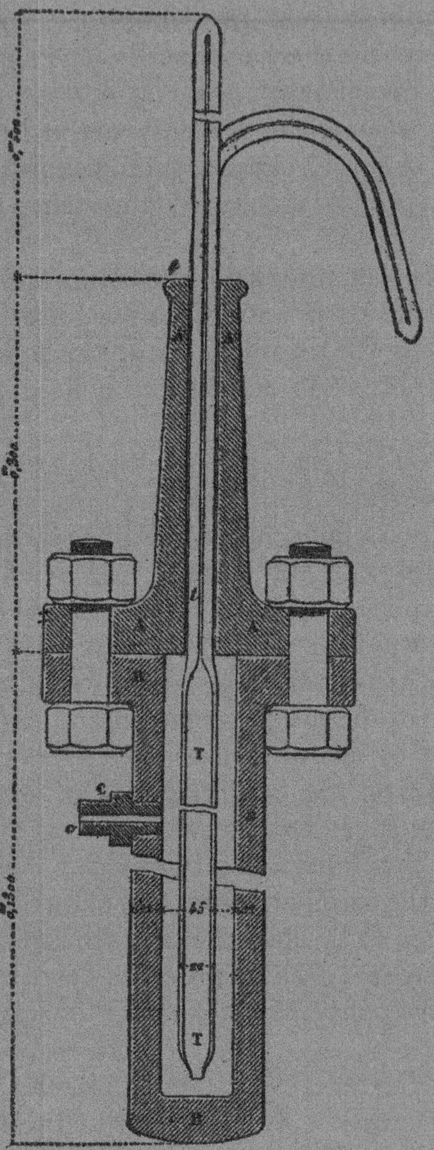

D. Colladon's Apparatus of 1828.

which rose from it was from 0·06 to 0·08 inch internal diameter. The steel reservoir was about one and three-quarter inches in internal diameter and about five and a half inches high.

A steel reservoir, *B*, held mercury. By a tube, *C*, connection could be made with its interior. An extension, *A A*, was bolted firmly to it. A glass gas tube, *T T*, open at the bottom, with a long capillary tube, *t t*, rising from its upper end, was mounted and inclosed in the mercury cistern as shown. The end of the capillary tube was closed. A limited amount of mercury only was required, as water could be used above it in the reservoir, *B*.

The capillary tube nearly fitted the long tube through which it passed, and the joint between metal and glass was made secure by gum lac.

Colladon worked at —30° C. (—22° F.) and 400 atmospheres pressure, without result.

The principal parts of his apparatus are still in existence, carefully preserved in the offices of the Société genevoise pour la construction des instruments de physique, in Geneva, Switzerland. An accurate sectional view of the same, reproduced here, is given in the *Annales de Chimie et de Physique*, fifth series, vol. xiii., plate facing page 288.

Thilorier applied the pressure produced in the generation of a gas to its own liquefaction. His pattern in this was Faraday, and he has been followed by Pictet and some others. He worked upon carbon dioxide gas, the gas familiar to all as the one which escapes from effervescing liquids. A pair of cast iron vessels were employed. In one the gas was generated, in the other it was received after generation, and the pressure alone was relied on to produce the liquefaction. He had no idea of applying refrigeration.

Producing liquid carbon dioxide on the large scale, he found, on releasing it from pressure, that the now familiar solid carbon dioxide was produced in snow-like masses. This gives an admirable example of the cold of a boiling liquid. The liquefied gas boils so energetically that it renders a quantity of heat "latent," or uses up heat energy, and the chilling of it is so great that some of it becomes a solid.

When Thilorier first observed this, he attributed it to the moisture of the air, and thought that the white solid was snow. A committee of the French Academy of Science examined it and found that it was carbon dioxide gas solidified.

The original Thilorier apparatus for liquefying carbon dioxide was made of cast iron, as has just been stated. In 1835 one of them blew up at the Ecole de Pharmacie, Paris, and tore off both legs of the unfortunate operator, M. Hervy. The use of the cast iron apparatus was proscribed on account of this accident. Mareska and Donny then modified the apparatus by constructing it without employing cast iron for generator or receiver.

In Liebig's chemical letters is given an account of this accident, which, as expressed in the words of the great German chemist and writer, is worth quoting in full:

"A melancholy accident occurred at Paris which proved the extreme danger of the preparation of liquid carbonic acid by the action of sulphuric acid on bicarbonate of soda, which is accompanied by a strong disengagement of heat. Just before the commencement of the lecture in the laboratory of the Polytechnic School, a cast iron cylinder two feet and a half long and one foot in diameter, in which carbonic acid had been developed for experiment before the class, burst, and its fragments were scattered about with the most tremendous force; it cut off both the legs of the assistant, and the injury was followed by his death. We can scarcely think without shuddering of the dreadful calamity which the explosion of this vessel, formed of the strongest cast iron and shaped like a cannon, would have occasioned in a hall filled with spectators, and yet the apparatus had been often used for the same experiments, which naturally banished all idea of danger." (Liebig's "Familiar Letters on Chemistry," London, 1851; letter x., pages 130, 131.)

The Thilorier improved apparatus is shown in the

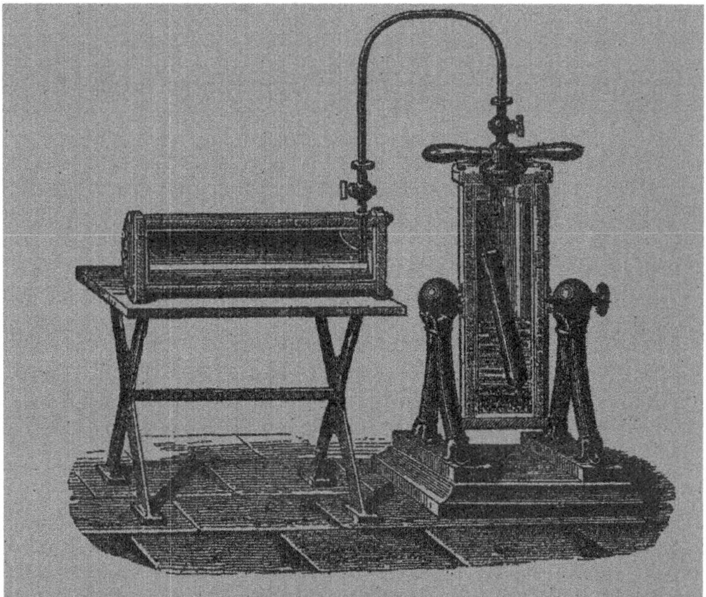

Thilorier's Apparatus for Liquefying Carbon Dioxide.

illustration. The right hand vessel, carried by trunnions, is of lead, inclosed in copper, with iron hoops or bands. It is connected by a tube with screw joints and connections as shown to a cylindrical receiver of similar construction. The tube connecting the two is of copper and has two stopcocks. The size of the apparatus as made may be gauged from the fact that the generator was of 6 to 7 liters capacity, or nearly 2 gallons.

To use the apparatus, the two vessels were first disconnected. Eighteen hundred grammes of sodium hydrogen carbonate (common baking soda), with four liters of water, were placed in the generator. A cylindrical vessel containing one thousand grammes of sulphuric acid was placed in it. In some constructions this vessel had a wire-like projection from the bottom designed to keep it in position, as shown in the cut.

The generator, which is the left hand vessel in the cut, was closed, the top being screwed on and the cock closed. By rocking and inclining it, the acid was discharged with some degree of control upon the sodium carbonate solution and upon the undissolved salt, and the gas was produced.

The generator and receiver (the right hand vessel) were now connected by the copper tube, the cocks were opened, and the gas rushed over into the receiver. A minute of time was allowed for the establishment of equilibrium, the faucets were closed, and the vessels were again disconnected.

The residual gas in the generator was blown off, the top was removed, and the whole operation as described was repeated. Five to seven repetitions were required to produce four liters, or a little over a gallon, of liquid carbon dioxide.

It is calculated that, with the apparatus charged as described, there was room in the generator for about one liter or a quart of gas. At the temperature of 40° C. (104° F.) the pressure would rise to one hundred atmospheres.

Thilorier did some good work on liquid carbon dioxide. As far back as 1835 we find a paper of his in the *Annales de Chimie et de Physique* on the properties of the liquefied gas. The extraordinarily high expansion of the liquid is spoken of, and the figures as he determined them are given. He finds it insoluble in water and in fatty oils. He gives a freezing mixture based on its employment, suggesting a mixture of liquid carbon dioxide and ether.

He found that this gave a frigorific agent of great power. By placing liquid carbon dioxide in a vessel provided with an outlet in the form of a blowpipe jet, he was able to produce local cooling effects. A jet of vapor would rush out, and would have great chilling powers. The arrangement he terms a *chalumeau de froid*—a cold blast blowpipe. He hopes for still better effects from a mixture of carbon disulphide and liquid carbon dioxide.

It is interesting, forty to fifty years later, to find the idea of producing cold by a jet from a liquefied gas again brought forward. Cailletet proposed to utilize the latent heat of liquid ethylene in this manner. The subject will be found treated on page 198 of this work, and Dewar used an escaping jet of liquefied hydrogen to freeze air and oxygen into solid white, icelike masses, as described on page 269.

Eleven years after Thilorier had devised his dangerous apparatus, a new one was produced by an Austrian scientist, which apparatus was comparatively safe. In a pump was the compressor, and a relatively small receiver, artificially cooled, took the place of Thilorier's large vessel. It was in 1845 that the apparatus was produced, and subsequent changes materially improved it. Johann Natterer, of Vienna, was its originator.

The apparatus consisted of a vertical compression pump actuated by a crank with flywheel. The pump was mounted in an inverted position and delivered the gases which it compressed upward from its highest point, which in its inverted position was really the bottom of the pump barrel. It was surmounted by a wrought iron reservoir of about one liter capacity which was strong enough to withstand a pressure of 600 atmospheres.

The liquid gas reservoir, slightly pear shaped, was surrounded with a basin of copper, designed to hold a cooling mixture. The pump had a solid piston. At the point where the pump barrel connected with the reservoir was a valve which opened upward. The gas to be liquefied was conducted to the lower end of the pump barrel, where a tube entered it far enough from the end to be above the solid piston as it reached its lowest point of descent.

An important modification was introduced by Bianchi. He surrounded the pump barrel with a jacket of metal, and let the liquid which drained from the refrigerating basin flow down and fill this jacket, whence it could be drawn from time to time by an outlet cock. Thus the pump barrel was cooled and a better working insured as regards the lubrication. The compression of a gas produces heat, and this dries up most lubricants. The gas also was thus delivered at a lower temperature to the reservoir, which in itself was an advantage. Those who have used compression pumps are familiar with the heating effect, which can be observed even in a bicycle tire pump when inflating a pneumatic tire. A second jacket surrounded the piston rod, which jacket received the melted material flowing from the refrigerating basin, so as to cool the piston rod directly.

The liquid gas reservoir could be unscrewed from the pump and carried about. The valve at its base closed and prevented any escape of gas. At its top the reservoir was provided with a cock by which the gas could be drawn off.

The gas is made in a generator, and may be first introduced into india rubber bags, which supply it to the apparatus, as shown in the cut. The drying apparatus, shown in the cut, is a Wolf's bottle (three-necked bottle) charged with a drying agent. The drying agent may be sulphuric acid, chloride of calcium, or some other of the regular materials used by chemists to remove water from gases.

The apparatus, which is a sort of classic, shows every sign of being designed to insure perfection rather from a mechanical than scientific standpoint. Simply for lecture demonstrations it is rather effectual, and is considered safe—something which cannot be said of some of its predecessors. The early experimenters, from Northmore down, have been troubled by explosions which culminated in the killing of a man, as already alluded to in the description of Thilorier's apparatus.

In the cut the entire apparatus is shown mounted and ready for work, and a sectional view on a larger scale shows the interior of the pump, gas reservoir and connections. *A* is the liquid gas reservoir, with its escape valve, *r*, *x*, for drawing off the liquid, and its self-acting base valve, *S*, through which the gas enters. It will be seen that, to draw off liquefied gas, the reservoir must be inverted. *B* is the cool-

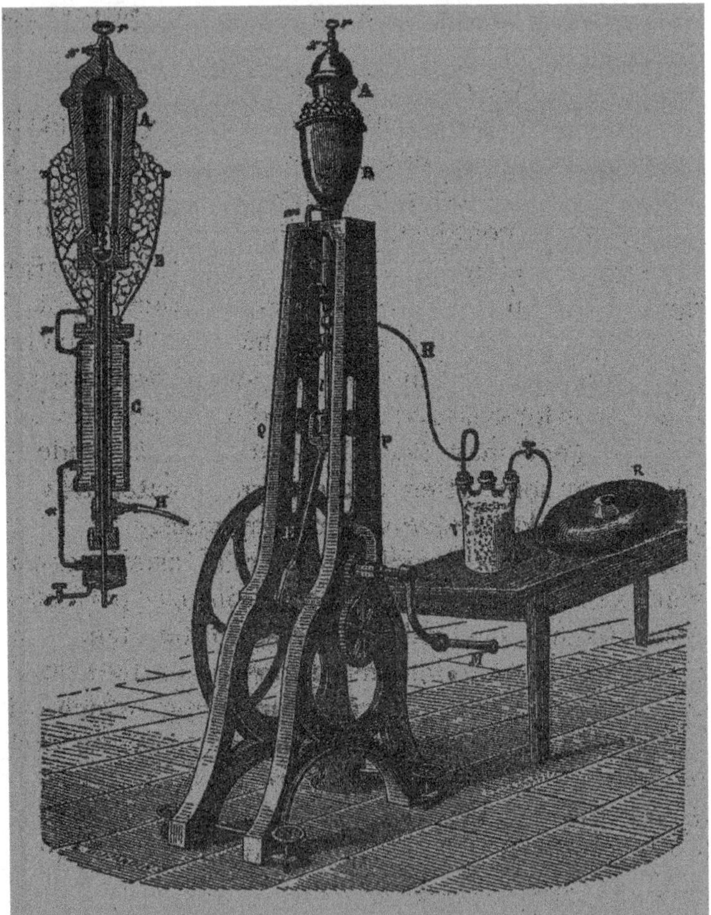

Natterer's Apparatus for Liquefying Carbon Dioxide.

ing basin; *m*, *n*, *o*, the drainage pipes and cock; *C*, the cylinder or pump barrel cooling jacket; and below is seen the small jacket for cooling the piston rod. The piston rod, *t*, with pitman, *E*, works in the slides, *P*, *Q*, in the massive metal frame. The gas from the bag, *R*, dried in its passage through the bottle, *V*, enters by the pipe, *H*.

It is Natterer who is celebrated for his liquefaction of nitrous oxide gas on the large scale, and who mixed the liquid with bisulphide of carbon for the production of an intense yet manageable refrigerating agent for scientific uses.

For determination of low temperatures Natterer used a thermometer filled with phosphorus chloride. This he told orally to Prof. Wroblewski or Olszewski (*Wiedemann's Annalen*, 1883).

The old apparatus was quite troublesome to use. It required one to one and a half hours' intermittent pumping to complete the operation. The piston rod had a way of heating, and this interfered with its lubrication; so that the operator had to stop from time to time to oil it, and this gave it a chance to cool.

When the receiver was two-thirds full 450 grammes of liquid carbon dioxide could be taken from it.

In the Leipzig *Journal fuer praktische Chemie* for 1845 is to be found a description of the early form of Natterer's apparatus, unimproved by the auxiliary cooling jackets shown in the more modern apparatus illustrated by us. The article is by Prof. Pleischl, and is quite quaintly expressed, or at least reads so in the light of over half a century's developments.

Prof. Pleischl notes the danger incident to the use of Thilorier's apparatus, and speaks of the death of Hervy, who was killed by its explosion some years previously. He says that his talented young student

Johann Natterer had succeeded in liquefying carbon dioxide with an air pump, and that led to the construction of what is known as Natterer's apparatus. The great safety of the new pumping system is quite enthusiastically commented on, and more notes of a public exhibition given on March 11 are embodied, at which exhibition carbon dioxide snow produced by Natterer's process was shown to a delighted audience. It was mixed with ether and used to freeze mercury, among other experiments.

Natterer made great efforts to liquefy the more permanent gases, but without success, and seems to have greatly regretted that better fortune did not attend his work. He carried his pressures up to nearly 4,000 atmospheres, or double the pressure produced in a cannon by the exploding powder. Some of his work is described in the *Wiener Berichte*, vols. v., vi. and xii. A rather complicated screw pressure apparatus is described and illustrated, by means of which he performed his high pressure experiments and determined quantities of data of the compression of gases under pressure. In vol. xii. of the *Berichte* he expresses his regret at not succeeding in liquefying gases.

Had he given the same attention to cooling his gases that he did to compressing them, he might have had a different tale to tell. The realization of all that the critical temperature means has given the liquefaction of gases its new aspect, and has led to the recent triumphs.

Far too little attention is given to Natterer's excellent work. He subjected gases under perfect control and visibility to most enormous pressures, and certainly to that extent helped to prove the doctrine of the critical temperature.

In 1888 Amagat, carrying pressures up to 3,000 atmospheres, got some discrepancies in his compression figures as compared with those of Natterer.

Andrews' Apparatus for Compressing Gases.

The work done by Thomas Andrews, of Belfast, in 1861 to 1870, as determining the existence of a critical state, is classic, and his simple apparatus is shown in the cut. A small glass tube contains the gas; a short column of mercury closes the tube below the gas; the upper end of the tube is sealed. The tube passes through a brass block, *E*, which is held by screw bolts on the end of a copper tube, *R*. A perforated block with screw thread cut in the perforation closes the lower end, and a steel screw, *S*, passes through the hole and closes it. All is packed so as to secure absolute tightness. The copper tube is filled with water. On screwing in the steel screw, the water is forced up against the mercury in the glass tube, *g*, and the mercury, in its turn, is forced up and the gas is reduced in volume, the object of the mercury being to cut off the water so that there shall be no action of the water on the gas.

To use mercury in this way, the tube, *g*, has to be of small caliber, or else the mercury would drop out. But another reason obtains. The steel screw is small, and the tube must be of the volume, or not much in excess of the volume, of the portion of the screw which can be screwed in and out.

By screwing in the screw the pressure could be raised to 500 atmospheres. Sometimes the tube was bent downward, so that its end could be placed in a freezing mixture, as shown in Colladon's apparatus, page 136.

Other varieties of the apparatus are shown in his paper published in the *Transactions* of the Royal Society of England for 1869.

In his early work he had used the compression produced by the electrolysis of water. If two terminals of an electric circuit of about two volts or more difference of potential are placed in a vessel of acidulated water, or of a solution of various chemicals, such as sodium hydrate or potassium hydrate, gaseous oxygen will be liberated from one terminal and gaseous hydrogen from the other.

The illustration shows a simple arrangement for carrying out the experiment. In the background is seen the battery. In the foreground is the decomposition vessel, with two spiral terminals or electrodes immersed in it. Only the spiral ends of the electrodes are bare. The other parts are covered by a tube of india rubber. The bare ends are inclosed in inverted test tubes, themselves filled with the solution. When the battery is connected as shown, bubbles rise from the wires, and hydrogen and oxygen gases collect in the test tubes.

Now, if such electrodes with some solution were introduced into a hermetically sealed and very strong vessel, the two gases would be evolved and enormous pressures could be generated by the quiet effects of the electric current. This is what Andrews did in his early work. With such apparatus he reduced oxygen gas to one three-hundredth of its volume.

In his later work, using apparatus on the principle

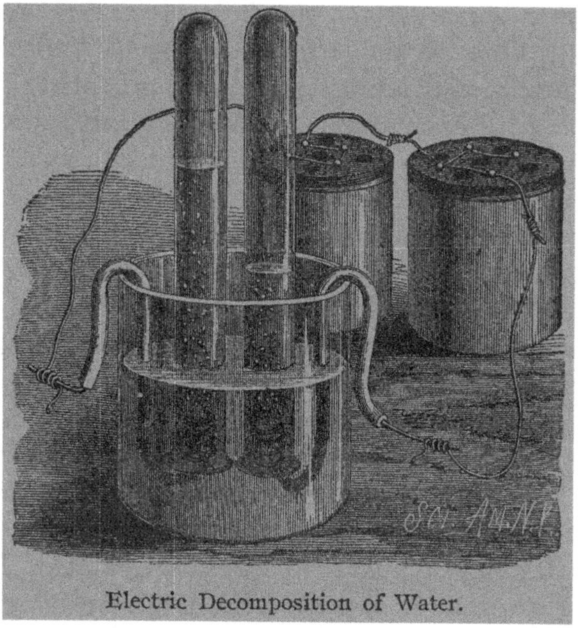

Electric Decomposition of Water.

described above, and using strong capillary glass tubes for the compressed gas, supplementing high pressure by cold of —106° C. (—159° F.), he reduced air to one six-hundred and sixty-fifth part of its volume. He got no result with any of what he called the six non condensible gases.

These were hydrogen, oxygen, nitrogen, carbonic oxide, nitrogen dioxide and marsh gas.

One of Andrews' principal papers, utilized above, is published in the *Transactions* of the Royal Society, as quoted, with very elegant cuts of the apparatus. It appears in a translation in the *Annales de Chimie et de Physique* of 1870.

Clerk Maxwell was much interested in Andrews' work. One of his letters alluding to Andrews' experiments, and addressed to the scientist, and bestowing his encomiums on his explorations into the realm of gases, is given in Tait and Brown's memoir on the life of Andrews.

The date of Andrews' work is generally put about 1862, one of his principal papers being published twelve years after his researches were made.

We have now reached a period whose history demands a somewhat different treatment. Up to the last date mentioned certain gases had resisted all attempts at liquefaction. Those which had been liquefied had been the subject of experiments on the small scale, and the efforts of investigators had been directed to the attaining of purely theoretical results. Two investigators now appear who profoundly modified the views of the scientific world. Pictet and Cailletet demolished the old division of permanent gases, and in doing so had a close race for priority. The French scientist Cailletet was awarded the priority by a few days only. But the work of the two men was so different in its scope and results that they should be considered hardly as rivals. Cailletet, by accident, produced mists in a small glass tube. These mists were due to the momentary liquefaction or reduction to the vesicular state of the gases contained. Pictet, on the other hand, directed his efforts from the first to producing a tangible quantity of liquefied gas. He was the first to secure this result; he was the first to produce a jet of liquid oxygen; he established the system of cascade or closed cycle refrigeration that has been the guiding principle for some twenty years of laborious investigation. Basing his work on Pictet's cycles, Dewar filled the Royal Institution laboratory with machinery and produced liquid gases by the gallon. Wroblewski and Olszewski combined Colladon's and Cailletet's methods with Pictet's cycles for the attainment of their results.

It should be felt that Pictet and Cailletet are to be placed side by side, and that no question of priority should be appealed to as existing between them.

Raoul Pictet

CHAPTER VIII.

RAOUL PICTET.

The life of Raoul Pictet—His education—His ice machines—Disputed priority—Honors awarded—His apparatus for liquefying gases—Description of its operation—Temperatures of the cycles of operation—His dispatch of December 22, 1877, to the French Academy—Regnault's statement—Hydrogen—His dispatch of January 11, 1878, to the French Academy—Olszewski's comments on the hydrogen experiment—Pictet's arrangement of pumps—His desire to produce liquid oxygen in quantity—Comments on his work—The *liquide Pictet.*

Raoul Pictet was born in Geneva, Switzerland, on August 4, 1846. He finished his studies in the Academy of Geneva when eighteen years old, and published some memoirs on binocular vision and on the resistance of the air. He went to Paris, and, although a foreigner, was received as a student at the École Polytechnique in that city. He also took courses in the College of France, and in the Sorbonne. There the young student became the friend of the greatest French scientists, Wurtz, J. B. Dumas, Regnault, Quatrefages and others. He received recognition from a most distant quarter, when the Saint Petersburg Academy of Sciences crowned his investigations of binocular vision and offered to publish all of his researches in full.

Three years were devoted in great part to the

study of thermodynamics. He made during the interval several long tours, and then returned to Geneva.

At the age of twenty-five he enterea the service of the Viceroy of Egypt. He was charged with establishing a course of instruction in experimental physics at the Ecole Superieure, in Cairo. While thus occupied he gave a good example of his aptitude for languages, acquiring Arabic in a few months' study.

Three years were passed in Egypt, and his life there gave rise to various interesting memoirs. The atmospheric phenomena of the desert, solar action, dust, whirlwinds and eddies, the temperature and floods of the Nile, were among the subjects studied and written on. He organized hunting expeditions into the interior, enriching with the spoils the museums of Cairo and of Naples.

The poisonous reptiles of the Nile regions, one of whose ancestors may be assumed to have inflicted the death wound on Cleopatra, attracted his attention, with a view to combating their venom in the human system. He collected snakes, and studied their poison in its action on the animal system. At one time he had four hundred specimens of Nile snakes in captivity. The natives of the region, it is said, still speak of the Geneva scientist who strove to diminish the deaths due to serpents' bites.

In 1877 Geneva claimed her son, and he accepted there a chair of physics and mathematics in the University of Geneva. He had for some years made ice machines, and had invented a process for freezing large areas of ice for skating, being a skater of no mean order himself. London, Manchester and other places saw skating rinks constructed on the Pictet system.

On his establishing himself once more in his native city, he was well prepared to begin his work on the liquefaction of gases. His work is detailed elsewhere. His friend Prof. Dufour, of the University of Lausanne, describes a visit made by special invitation to the buildings of the Société genevois pour la construction des instruments de physique. The visitors were a number of professors and scientists from Lausanne, the date was December 29, 1877, and Pictet showed them the liquefaction of oxygen.

It will be seen that his early work in the production of low temperatures was in the practical line, and, therefore, on the large scale. This it was which gave his liquefaction of gases such value. He was not content to produce an infinitesimal amount of liquid. The desire to produce tangible quantities was ever present in his mind. As regards the method, it was based on practically successful processes. The engineer's mind appeared in the working of his cumulative cold-producing circuits, and he established a system which has done service for over twenty years of investigation in England, Holland, Poland and Germany.

As will be seen by those who follow the dates given in this book, there was a close coincidence between the dates of Pictet's and of Cailletet's liquefactions of oxygen. This was the origin of hot disputes waged by the political dailies, for in Europe all sorts of pretenses are seized upon for political effect The methods followed and apparatus employed by the two scientists were so radically different that at last Cailletet protested, the war ceased, and an intimate friendship was formed between the rivals that was never broken. Regnault interested himself in the work he had so long followed, and informed the Academy of France that Pictet's system of cumulative cold-producing circuits, to his knowledge, dated back five years, and that the experiments might have been performed five years earlier had events favored the work.

The dispute was ended, and Pictet received the decoration of the Legion of Honor. France, as always, was generous to the foreign rival for scientific honors.

The mechanical theory of heat was, about this time, investigated by him in union with M. Gustave Cellérier for eighteen months. The study was so intense that Pictet nearly broke down in health on the completion of the work.

In 1878 he received from the International Exposition at Paris the gold medal, and in the same year the Royal Institution of England gave him the Davy medal.

In 1880 he went to Berlin, and there established a low temperature laboratory. The study of frigotherapy was taken up, and the purification of chemicals by intense cold was worked upon.

The cities of Antwerp and of Rome have recently honored him by diploma and medal of honor. In 1895, the Société Industrielle du Nord de la France gave him its grand medal of honor at Lille.

His life has been written from the standpoint of a dear friend by Prof. Henri Dufour, of the University of Lausanne, Switzerland. To him the author of this book is indebted for copious notes on the life of Pictet, and interesting accounts of the personal traits of the distinguished scientist, who knows how to charm children by feats of legerdemain as well as to interest and delight the world of scientists by his achievements in physics and in the realm of low temperature.

He has entered the field of intellectual and moral philosophy in his treatise entitled *Etude Critique du Matérialisme et du Spiritualisme par la Physique Expérimentale.* This is a large octavo, and investigates the relation of material energy and mental operations most interestingly, and is a scientific protest against doctrines leading to the depression or despair which sometimes seems to obtain a foothold among scientific students.

Pictet's apparatus by which he succeeded in liquefying oxygen is described in the *Comptes Rendus,* vol. lxxxv., page 1214. The illustration we give is substantially identical with the one given in the *Comptes Rendus,* except that it is completed by the introduction of the gas burner for heating the oxygen retort, and that a manometer or pressure gauge and outlet cock are shown at *R, N.*

The Pictet apparatus, as shown, deserves especial attention because it is the original of a type which only now encounters in self-intensive processes a really efficient rival. It was far in advance of its time. Apparatus of its type was added to the Colladon apparatus by Wroblewski and Olszewski for their work. Dewar employed it in his Royal Institution researches, and the extensive apparatus in the Leyden University cryogenic laboratory is based upon its lines. This apparatus was the first to produce a stream of liquid oxygen, and it cannot be awarded too high a place in the history of low temperature experimentation and research.

L is a wrought iron retort calculated in the original Pictet apparatus to resist 500 atmospheres pressure. Subsequently, it is said to have been made

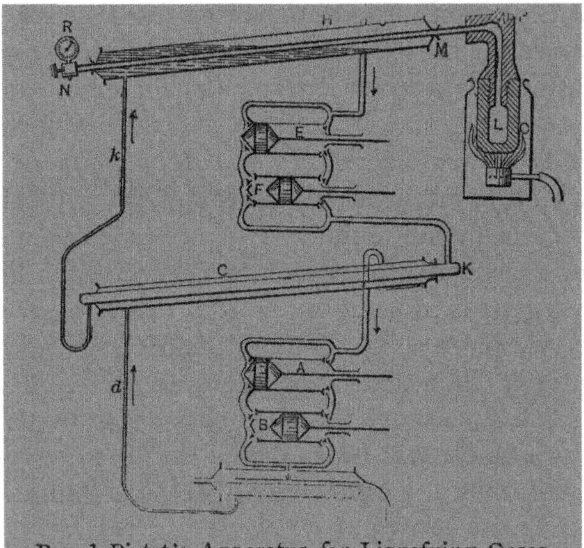

Raoul Pictet's Apparatus for Liquefying Gases.

stronger, so as to be able to withstand three times this pressure. A weighed amount of potassium chlorate was introduced by the opening, *P*, which was then closed. On heating it by the lamp, *O*, the quantity of oxygen to give any desired pressure was produced, such quantity being determined by the weight of potassium chlorate employed.

The tube, *M*, was thus filled with oxygen at a pressure regulated by the weight of potassium chlorate. Pressure was thus produced, which is one element of the process of liquefaction. The next step is the cooling of the compressed gas.

The condenser jacket, *C*, contains liquid sulphur dioxide. This tends to evaporate and to produce thereby great refrigeration. From the upper end of the jacket, *C*, a pipe goes to the pumps, *A* and *B*. These pump out gaseous sulphur dioxide. The liquid sulphurous oxide in *C* boils, therefore, with greater rapidity than ever, and produces greater cold. The gas goes through the pumps and is compressed by them in the condenser jacket, *D*. The outlet of this condenser jacket, *D*, is a narrow pipe, *d*, which, being of small diameter, produces the requisite pressure to condense the sulphur dioxide to a liquid. Through the condenser jacket, *D*, a pipe runs, and cold water passing through this pipe cools the sulphur dioxide as it comes heated by compression from the pumps, *A* and *B*.

The upper system of pumps and cooling arrangements is almost in exact duplication of what has just been described, except that liquid carbon dioxide takes the place of liquid sulphur dioxide, and the liquid sulphur dioxide under exhaustion takes the place of the cold water.

The condenser jacket, *H*, contains liquid or solid carbon dioxide, which constantly evaporates. The pumps, *E* and *F*, pump gaseous carbon dioxide out of the upper end of *H* and condense it in the tube, *K*, where it is cooled by the boiling sulphur dioxide. The small pipe, *k*, creates the requisite back pressure for the liquefaction, and a constant circulation is thus maintained, and the boiling carbon dioxide keeps the tube, *M*, inclosed in the condenser jacket, *H*, at a very low temperature.

The following figures are given in the *Comptes Rendus* as the data of the first successful attempts at liquefying oxygen:

The sulphur dioxide liquefied in *D* at a pressure of two and three-quarters atmospheres, and produced by its evaporation in the jacket, *C*, a temperature of —25° C. (—13° F.) The carbon dioxide liquefied in *C* at a pressure of five atmospheres and a temperature of —65° C. (—85° F.) The tube, *M*, by the evaporation of the cold carbon dioxide, was kept at a temperature of —140 C. (—220 F.)

In the improved apparatus, the tube, *M*, was made of copper, and the liquefied gas was withdrawn at *N;* but in the apparatus of 1877, as shown in the *Comptes Rendus*, the tube in question was unprovided with a faucet, and its lower end was within the condenser jacket, *H*. The tube was one meter or a little over a yard long.

In the original experiments, which now may be considered historic, the pumps were worked for several hours circulating the sulphur dioxide and carbon dioxide. A 15 horse power engine was employed to drive them. Meanwhile oxygen was being evolved, and the pressure was brought up to 320 atmospheres. Then the cock at *P* was suddenly opened, and the sudden expansion of the tremendously compressed and very cold oxygen absorbed so much heat energy, rendering the heat latent, that the temperature fell still further, the oxygen was liquefied in part, and the tube, *M*, was filled to one-third of its length with the liquid. The tube being of 1 centimeter (0·4 inch) internal diameter, it will be seen that this was a considerable quantity of oxygen—about 22 cubic centimeters or 1½ cubic inches of the liquid.

On inclining the tube by raising the lower end, the liquid rushed out of the orifice at *P* ("et jaillisse par l'orifice en inclinant l'appareil"). It will be remembered that in the original apparatus there was no way of opening the lower end of the tube, which was closed and within the condenser jacket, *H*.

Pictet's dispatch announcing the success of his experiment, on which so much time, thought and expense had been lavished, was received by the French Academy of Sciences on December 22, 1877, at 8 P. M. It was as follows:

"Oxygène liquéfié aujourd'hui sous 320 atmospheres et 140° de froid par acide sulphureux et carbonique accouplés.

"Signé,

"RAOUL PICTET."

(TRANSLATION.)

"Oxygen liquefied to-day under 320 atmospheres and 140° of cold by suphurous and carbonic acid working together.

"Signed,

"RAOUL PICTET."

The terms sulphurous acid and carbonic acid are synonyms for sulphur dioxide and carbon dioxide.

The substitution of the open copper tube *M* for the closed one and the use of the manometer, *R*, and cock, *N*, are later modifications. The temperature of the oxygen generating retort, *L*, is now put at 485° C.

Pictet's Gas Liquefaction Apparatus of 1877.

(905° F.) The manometer in the course of the condensation rose gradually until it indicated a pressure in *L* and *M* of 500 atmospheres. The gas began to liquefy and the pressure fell to about 320 atmospheres. On opening *N*, the oxygen rushed out under the great force of the pressure with violence, looking like a dazzling white pencil. The escape lasted for 3 or 4 seconds; the manometer showing some 400 atmospheres, which rose again and again fell when liquefaction occurred.

The large cut shows the general disposition of Pictet's apparatus as installed in Geneva.

F and *H* are two boxes packed with non-conducting material, and in each of these are two concentric tubes constituting a condenser of the Liebig type.

In *F* is the oxygen liquefaction tube surrounded with another tube through which the carbon dioxide, solid, liquefied and partly gaseous, circulates. This corresponds to *M* and *H* of the diagram on page 158.

In *H* is the carbon dioxide tube, where the gas from the outside tube in *F* is cooled by boiling sulphurous oxide, which is in a tube inclosing and concentric with the carbon dioxide tube. These are *K* and *C* of the diagram.

G is a gasholder filled with carbon dioxide gas. *K* is a reservoir of liquid sulphurous oxide. *P* are the pumps, and *B* is the oxygen retort.

A moment's inspection of the cut, after study of the cut on page 158, will suffice to give a full understanding of the operation of the apparatus.

It was no easy matter to obtain the small quantity of liquid gas that greeted Pictet's vision on the twenty-second of December, 1877—the first sight of liquid oxygen in quantity that ever was granted to man. Regnault told the French Academy that he had assisted Pictet and De la Rive five years before the date of the liquefaction in experiments on liquefying gases, and the work of five long years reached only then its culmination.

Pictet examined with a polariscope the escaping jet of liquid oxygen as it rushed violently out of his tube, and thought that he obtained evidences of the presence of solid particles in the stream.

Pictet did not rest here. The few cubic inches of liquid oxygen which he had produced acted as an incentive to go further, and he endeavored to liquefy hydrogen.

The details of the experiment are given in the *Comptes Rendus*, vol. lxxxvi. They are contained in a dispatch from Geneva, followed by a letter.

He wished to make his hydrogen by heating a solid substance in a retort, so as to preserve the general system of his oxygen method. Accordingly, he employed a mixture of potassium formiate and potassium hydrate. This mixture, he says, gives pure hydrogen, free from water or carbon dioxide, and leaves a non-volatile residue.

On applying heat to his retort, the pressure ran up to 650 atmospheres and then remained stationary. The temperature of the gas tube was about —14 ,° C. (—220° F.) Enough gas was generated to measure 252 liters at 0° C. (32° F.) The cock was opened and what is described as a steel blue jet escaped with a sharp hissing sound. A length of 12 cm. (about 5 inches) of the jet was opaque. The jet struck the floor with a sound like hail. The hissing sound changed its character until it resembled the noise produced when metallic sodium is thrown upon water. The pressure ran down to 370 atmospheres and the delivery became intermittent, the tube or cock being choked. For over fifteen minutes the delivery by the jet occurred in intermittent discharges.

The liquefaction of hydrogen has been felt to be open to doubt. The fact that the temperature as given is entirely insufficient, at any pressure, to cause liquefaction does not at all invalidate the experiment. The release from high pressure of the gas, bringing about its expansion, rendered heat practically latent and caused intense chilling of the gas, already at very low temperature, and might produce liquefaction of the hydrogen. The experiments of Cailletet confirm strongly this view of Pictet's experiment. But we know that no hydrogen was liquefied in volume in the tube before it was opened.

Ten years later Olszewski tried to throw some doubt on the method followed in the hydrogen experiment of Pictet. He published in the *Philosophical Magazine* for February, 1895, a long article giving a full account of his work of bygone years, in which he, with Wroblewski, produced liquefied gases. This article is a statement of Prof. Olszewski's part in liquefying gases and air. In the course of the article he criticises Pictet's hydrogen experiment, saying that hydrogen made as Pictet made it would be contaminated with water and carbon dioxide.

As a piston works in a pump cylinder, what is termed clearance occurs. This is the failure of the piston to expel everything from the cylinder. It is mechanically impossible to do this with steel or iron parts, as the piston cannot well be so accurately made as to just touch the cylinder on its completion of a stroke. Even if it could, the valve passages would be left.

As all gases are elastic by nature, it follows that, when a pump is caused to operate upon a gas, the clearance of the piston is a great obstacle to its operation. As the piston of a pump cannot absolutely touch the cylinder end at each stroke, some gas must always remain in the cylinder, and during certain conditions of tension and compression, when the suction is of high degree, and the delivery is against a high pressure, the piston may work back and forth without any result whatever. The gas remaining in the cylinder ends may be enough in amount to prevent any movement of the suction or inlet valve, or to admit other gas if it were opened, and not enough, on the other hand, to open the outlet valve, or, if it were opened, to go through it.

This difficulty, inherent in all ordinary piston air pumps, Pictet avoided by coupling his pumps two in a set. Thus, when one pump was aspirating from the cooler jacket or other source of gas, it was delivering, not against a high pressure, but into the suction pipe of the other pump. The other pump took this partly compressed gas through its suction pipe as delivered by the first and gave it its second compression.

By this arrangement the difficulties were suppressed and the four pumps working in sets of two each operated perfectly. They were driven by band wheels at from 80 to 100 revolutions per minute.

The temperature was determined by a formula which is deduced from the mechanical theory of heat applied to change of state. The formula can be found in the paper of Prof. Pictet as given in the *Annales de Chimie et de Physique*, Paris, fifth series, vol. xiii., or in the *Archives des Sciences Physiques et Naturelles*, Geneva, January 15, 1878.

It is most interesting in this paper, which is the definite and authoritative presentation of the experimenter's views to find the following passage. It must be remembered that the oxygen had been liquefied in an opaque tube, that it was withdrawn therefrom by the cock under enormous pressure, and that the sight of the jet, which lasted only three or four seconds, was the nearest approach to really seeing liquid oxygen which the definite experiment afforded. We quote the passage:

"We must try to render this liquid oxygen *visible* by condensing it in transparent apparatus. The problem is very complex, bristling with practical difficulties. We must avoid the condensed ice (*givre*, hoar frost) which instantly forms on cold surfaces, and impairs visibility; we must have tight joints with fragile material," etc.

Had Pictet foreseen the importance of the spheroidal state in its relations to the handling of liquefied gases, and could he have divined how greatly it would facilitate all operations with them, he would have seen the difficulty disappear in great part. But no human being could have imagined how greatly the maintenance of the spheroidal state was to affect the question.

The same desire to get oxygen in quantity is here

discernible which formed the inspiration for Wroblewski, Olszewski and Dewar. A scientist might be satisfied with Cailletet's mist or with Pictet's jet, but they were not. The desire to see oxygen and the other gases liquefied in volume has proved itself no mere idle dream, but a real, earnest and scientific longing. The effort and desire to satisfy this longing has led to the achievements commemorated in this volume.

The oxygen in five of Pictet's early experiments was evolved from a mixture of 700 grammes potassium chlorate and 300 grammes potassium chloride. This mixture may be taken as a typical one.

The hydrogen mixture used in his experiment of January 10, 1878, consisted of potassium formiate, 1261 grammes; potassium hydrate, 500 grammes.

The importance and value of Pictet's early work cannot be overestimated. His double cycle with continuous liquefaction of the gases in the two refrigerating cycles has been the instrument of the greatest successes in the hands of subsequent workers. All who worked upon this line in those early days overestimated the importance of pressure, but the keynote of Pictet's work was a very advanced refrigerating apparatus. The critical temperature is the great element in attaining success in liquefactions. It would have been but a small change to have compressed by mechanical means the gas to be liquefied. Had he done so, the effect would have been twofold.

He would have had more gas to be acted on. As his experiments were conducted, he had a very limited supply of gas, and on opening the cock of his apparatus it rushed out violently, and a fleeting glance of a second or two at the liquefied gas was all that it was in his power to obtain. But had he gone a single step further, and connected a third pair of pumps to the inner tube, *M*, of the gas condenser, there is every probability that he would have succeeded in his long-cherished wish much better. To him might have been granted the success claimed by Olszewski, of pouring for the first time liquefied oxygen or air from one vessel into another. But the work of Natterer and Andrews had its effect, and high pressure was striven for, and static air and oxygen remained for several years an unfulfilled hope and expectation.

Pictet, in the year 1885, devotes a paper to a new refrigerant, which has been named from him the *liquide Pictet*. It is still used by him for the production of low temperatures. The paper will be found in the *Comptes Rendus*, vol. c. He suggests that, for the production of low temperatures, a mixture of two or more volatile liquids may be employed. It has been aptly said that in mixing metals so as to produce new alloys the metallurgist is able to produce so many new metals. Each alloy may be taken as equivalent to a new metal. The properties of an alloy are not the average of the properties of its constituents. In specific gravity, electrical conductivity, thermal and other properties no average can be traced in many instances.

Pictet found that the case was the same with mixtures of liquefied gases, and in the paper in question discusses at some length the use of such liquids, which at relatively low temperatures separate into their components. He gives a table of the boiling points of different mixtures of carbon dioxide and sulphurous oxide, using molecular mixtures, or mixtures in which the proportions of the constituents stand in molecular proportion to each other.

He succeeded in producing liquids which boil anywhere from —71° C. (—95° F.) to —7·5° C. (—18·5 F.) But this range of selection open to the physicist is not the only advantage. There is a sort of recuperative or self-intensive action involved which makes the *liquide Pictet* peculiarly available.

At low pressures its evaporative power is augmented by its disposition to dissociate molecularly, or to separate into the two gases, carbon dioxide and sulphurous oxide. At high pressures a sort of chemical affinity of low order seems to come into play, and the two gases liquefy much more easily than they do when unmixed. It is easy to see how this phenomenon lightens the work of the pump used to condense them. On the exhaust side the action is aided by the dissociation tendency of the liquids evinced in their gasification. This lightens the work of the pump, as it does not have to draw so hard to cause rapid evaporation. This evaporation is the refrigerating action.

At high pressures the chemical affinity also helps the work of the pump; for, less power being required to liquefy them than otherwise, the pump has not got to develop the same pressure as it would otherwise. Hence its work is lightened on the pressure side also.

This peculiarity is brought out by a comparison of the *liquide Pictet* (formula CSO_4) with sulphurous oxide. At high temperatures the vapor tension of the *liquide Pictet* is higher than that of the sulphurous oxide. But on increasing the pressure and lowering the temperature the vapor tension increases in a less rapid ratio with the *liquide Pictet* than with the sulphurous oxide, and at a low enough point the sulphurous oxide shows the higher tension. In graphic terms the curves of tension and temperature relations cross each other.

All of Pictet's work cannot be given within the limits of this book. This chapter gives the summary of his original liquefaction of gases. But his practical mind sought fields for the utilization of his discoveries, and in subsequent chapters will be found described his application of low temperatures to treatment of disease and to the purification and production of chemical and technical products.

CHAPTER IX.

Louis-Paul Cailletet.

The life of L.-P. Cailletet—His education—Honors received—His modification of Colladon's apparatus—Accidental liquefaction of acetylene by release—Description of his apparatus—How the apparatus was filled—The full apparatus with hydraulic press—Liquefactions of nitrogen oxide—Of carbon monoxide and oxygen mixed—Liquefactions of the same separately—His letter of December 2, 1877, to the French Academy—Liquefaction of nitrogen—Of hydrogen—Rival claims of Cailletet and Pictet—Mercury stopper method—Manometers—Original methods of testing—Eiffel Tower manometer—Carbon dioxide experiments—Mercury pump—High pressure gas reservoir—Ethylene as a refrigerant—Closed cycle method—Accelerated evaporation—Electric conductivity at low temperatures—Comparison of thermometric methods—La Tour's experiment repeated.

Louis-Paul Cailletet was born in Châtillon-sur-Seine, in the Côte d'Or, France, on September 21, 1842. He studied at the Lycée Henri IV. and then entered the École des Mines, Paris. On finishing his course he returned to Châtillon-sur-Seine and soon was placed in charge of his father's iron works at that place.

He made many researches into the working of blast furnaces, the problems of combustion and of metallurgy. The occlusion of gases and the causes of explosion of iron while in the process of forging were also investigated, and a number of his papers were published in different scientific journals.

His investigations in the field of the compression and liquefaction of gases began about 1876, and reached their culmination in his liquefaction of oxygen and other "permanent gases" in 1877 and 1878. But he did not desert the subject, and for years after numerous papers by him in the *Comptes Rendus* attest his interest in it and his indefatigable powers of work.

Honors were given him for his work, of which we do not give the full list. It must suffice to say that he was elected a *correspondant* of the French Academy of Sciences December 17, 1877. On April 28, 1884, he received the prix Lacarze from the French Academy of Sciences, for the liquefaction of gases, the report coming from the following distinguished committee: Profs. Chevreul, Fremy, Wurtz, Cahours, Friedel, Berthelot, Dumas, Pasteur and Debray; and on May 26, 1884, he was elected *membre libre* of the Academy.

He had done much work upon the other subjects when he took up the action of gases under compression. At first he had no idea of liquefying the permanent gases, but he was a keen observer, and this led to his success.

Looking back at the work of his predecessors, he found that they had settled upon one type of compression apparatus, which rendered possible the subjection of a considerable body of gas to an enormous pressure, and that in a transparent tube.

He had adopted Colladon's well known compression apparatus (page 136) for the purpose of his investigations, but he connected to the hydraulic press by which it was operated a valve for sudden release of the compressed gas from pressure.

He builded better than he knew. His release method introduced a factor which produced intense cold in the gas, which cold brought about its liquefaction. The importance of the critical temperature may have been perfectly well known in 1877, but it was not so fully appreciated as now. Cailletet, almost by accident, came upon a method which enabled him to liquefy gases, simply because it lowered their temperature below the critical point. But when Cailletet first lowered the temperature in this way he did it without the least idea of liquefying a gas. The liquefaction was accidental, and was not even recognized as being what it was.

The authoritative statements of each step of Cailletet's work, published as soon as each step was completed, are given in the *Comptes Rendus* of the French Academy. He follows the custom of some other scientists by giving in another publication the *résumé* of his entire work up to the time when it was practically complete. A paper by him, with illustrations of his apparatus, is published in the *Annales de Chimie et de Physique*, 1878, which does this for his first work on the liquefaction of gases.

Pictet follows a like course, publishing specific papers in the *Comptes Rendus*, and following them with a general illustrated description in other publications.

The work of Cailletet on the liquefaction of gases begins with his work on acetylene. From the somewhat concise statements in the *Comptes Rendus* we may trace his work as originally published. But it will be better to invert the natural order a little and first present the more general view of his operations with the description of his apparatus, and then give a brief recapitulation of the more important *Comptes Rendus* articles.

Cailletet's original liquefactions seem to have been less satisfactory than Pictet's, as the proof depended on the production of a mist or fog of the liquefied gas. He compressed the gas which he was working on, cooled it, and then suddenly released it from pressure. The quick expansion absorbed heat, the temperature fell and he got the mist, which he describes by the word *brouillard.* We find here an indirect appeal to the critical temperature. He refrigerated the gas to such an extent by the sudden expansion that it fell far below the critical temperature.

The experiments were easily performed, and could be repeated over and over again upon the same portion of gas during the same day, so as to acquire force by reiterated success. The apparatus and its use were both simple, relatively speaking, and as demonstrations the experiments were accepted by scientists of absolutely the highest standing as satisfactory.

The compression apparatus will be recognized as a development of Colladon's and of Andrews' apparatus, which is illustrated and described elsewhere (pages 136 and 147). The cut shows the essential portion of Cailletet's apparatus as given by him in his article in the *Annales de Chimie et de Physique* of 1878, in which journal he describes his work in more detail, or, at least, in more popular style than in the *Comptes Rendus*. In the latter publication, under various dates, are published the somewhat condensed statements of the results of his work, but in the *Annales* a general view of the course of experimentation which led up to his final liquefactions of 1877 is given.

Referring to the cut, *B* represents a heavy steel cistern into which a glass vessel, *T*, dips, whose upper end forms a tube, *T P*. This is sealed at the top, *P*, and contains the perfectly dry and pure gas. It is sealed with an absolutely tight joint where it passes through the metal piece, *A*. A gland, *E*, screws down against the flange on the bottom of *A*, squeezing it against the packing shown under *A*. *M* is an open glass vessel which contains a cooling mixture if such is desired to be used, and a glass shade, *C*, covers the upper part simply as a matter of security. The darkly shaded part within *B* and *T* represents mercury; the lighter shaded portion in *B* is water. A cock serves to draw to draw off the refrigerating agent from *M*. *U* is a pipe joined by the coupling, *R E*, to the mercury vessel. *S* is the platform which supports the shade, *C*, and refrigerant vessel, *M*.

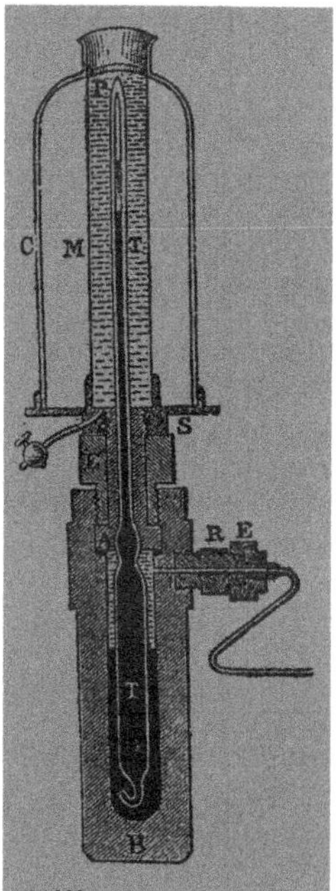

Cailletet's Liquefaction Apparatus.

When the apparatus is first set up, the level of the

mercury in *T* is much lower than is shown in the cut. It would be considerably below *T*, or not far from the bottom of the gas tube.

By a pump or hydraulic press water is forced into *B*. This forces the mercury up into the tube, *P T*, until the gas is greatly compressed. The upper portion of the gas tube, it will be seen from the construction, is the only part which is subjected to a bursting pressure, and it is so small in diameter that it can be made very strong without being of inordinate thickness.

The gas was compressed by a hydraulic press, as shown in the cut, page 180. A valve in the compressing press or pump was suddenly opened by the handle, *O*, and the gas was so cooled by its own expansion that a mist formed, which was composed of particles of the liquefied gas. The liquefaction consisted in the production of this mist.

In his original work Cailletet used a very powerful screw press worked by handles on a large fly-wheel. In the illustration of the entire apparatus the disproportion between the great compressing press and the little glass tube holding its minute quantity of gas is impressive.

The filling of the gas tube with dry gas was thus effected. The upper end of the tube was left open. A drop of mercury was placed in the large gas tube or bulb, the tube being held horizontally, and a tube from the gas evolution apparatus was slipped over the other end. A current of gas to be experimented on, purified by proper chemicals, was passed through the tube, and while it was still passing the upper end, *P*, was sealed tight with the blowpipe or blast-lamp.

This was done with the tube in an approximately horizontal position. Next the tube was returned into the vertical position with the sealed end uppermost. The drop of mercury ran down into the bent-up lower end, and the gas was thus hermetically sealed in the tube. It was then lowered into the reservoir of mercury, *B*. The connections were made and all was ready for the experiment.

The gas tube, it will be observed, differed from Colladon's in its bent-up lower end. This feature enabled the globule of mercury to act as a valve and seal the gas up in the tube before the latter was inserted in the cistern.

It is impressive to contrast the diminutive size of the liquefaction apparatus with that of the hydraulic press. The whole mechanism, whose size can be judged from the figure of the operative, is devoted to producing liquefaction phenomena in a glass tube of a fraction of an inch in internal diameter. The old error was perpetrated of overestimating the importance of pressure and underestimating the influence of reduction of temperature.

The first cloud he ever produced with a gas in his apparatus was with acetylene on sudden release from pressure, and it was unintentionally produced. He was experimenting with the gas, subjecting it to pressure not sufficient to liquefy it. He opened his release cock, and, as the gas expanded suddenly, he saw a mist or cloud form within the gas tube.

The first stroke of the piston of an air pump in exhausting a glass receiver produces such a cloud within the receiver, owing to the precipitation of moisture in the air by the cold due to rarefaction of

Cailletet's Complete Gas Compressing and Liquefaction Apparatus of 1877.

the air in the receiver. The appearance is very familiar to all who have used the old-fashioned air pump. It was, therefore, quite natural for Cailletet to conclude that the acetylene with which he was working was impure. He wished to avoid the presence of impurities. So he procured some absolutely pure acetylene gas from Berthelot's laboratory, filled his tube with it, and on compression and sudden release got the same cloud as before. He tried nitrogen dioxide and again the cloud appeared.

He now recognized fully what was occurring, and saw a very simple and effective way of showing the liquefaction of gases. He tried his famous experiment of December 2, 1877, in which he used oxygen gas and got the same appearance of a mist with it.

The large illustration shows the full apparatus used by Cailletet. *A* is a steel cylinder with plunger actuated by a screw, *F*, and held in brackets, *B B*. *M* is a wheel by which the screw is turned. The cylinder is filled with water by the glass funnel, *G*. To relieve the pressure when it might be desirable, a special valve operated by a wheel, *O*, was provided, and it was this valve which constituted the distinguishing feature of Cailletet's process and apparatus.

At *S* is a cross-connection to bring into connection the hydraulic cylinder, *A*, the liquefaction apparatus of page 177, and the gauges. Two of these are shown. One, designated by *N*, is a Thomasset manometer; *N'* is a Cailletet glass bulb manometer, such as spoken of on page 187.

The liquefaction apparatus, *m*, stands upon a shelf of iron, *p*, with set screws, *d d*, to secure the mercury reservoir, *a*.

The value of his work depended on the sudden release of the gas from pressure. As this was effected by opening a valve on the compressing apparatus, it caused the mercury to suddenly fall in the gas tube, but there was no loss of gas. The same sample of gas could be experimented with over and over again.

The sudden release, Cailletet calculates by Poisson's formula, should give a lowering of temperature of 200° C. (360° F.) This release constitutes the advance in his work over all his predecessors. As a physical demonstration, it gives a very elegant method of cooling a gas below its critical temperature. It is so direct an attack upon the molecules, and is so quick, as to effect the refrigeration without need of jacketing the tube. The expansion is almost perfectly adiabatic.

The pressure applied to the gas was determined by various manometers. One of his own devising, which we describe from papers in the *Comptes Rendus*, was employed, as well as another one by Thomasset. Both were connected to his compressing press. For lower pressures he could use an open end manometer of his own construction. This, however, was more adapted for standardizing purposes; his glass compression manometer was the instrument best adapted for use on his gas liquefying apparatus.

In the *Comptes Rendus* of October 29, 1877, page 851, he describes his work with acetylene. At 18° C. (64·4° F.), and a pressure of 83 atmospheres, he got drops of liquid acetylene. Then, on suddenly releasing it from pressure, a fog or cloud of acetylene formed. He reports the liquid as colorless, mobile and of a high refracting power.

A letter from him is given in the *Comptes Rendus* for November, 1877, page 1017, in which he says that he has liquefied nitrogen dioxide, using a temperature of —11° C. (12·2° F.) and a pressure of 104 atmospheres. At 18° C. (64·4° F.) it resisted a compression due to 270 atmospheres. Formene was tried, and on release gave a mist.

Next, in the same volume, he says he got a mist with a mixture of carbon monoxide and oxygen, and we find in the same volume, page 1217, his letter to the French Academy of Sciences, announcing the liquefaction of oxygen and carbon monoxide. It is dated December 2, 1877, and is given below. The very modest tone of the letter, and the feeling of the writer that his mist of condensed gas was hardly a sufficient liquefaction, are very evident, and inspire the readers of the letter with additional confidence in Cailletet's work.

The letter is historic, as it is used to determine the question of priority between the French and the Swiss scientists, Cailletet and Pictet.

We give a translation of the letter:

"I hasten to tell you, you first, and without losing a moment, that I have liquefied to-day both carbon monoxide and oxygen.

"I am, perhaps, wrong in saying liquefied, for at the temperature obtained by the evaporation of sulphurous acid, say —29° and 200 atmospheres, I do not see the liquid, but a mist so dense that I can infer the presence of a vapor very near to its point of liquefaction.

"I write to-day to M. Deleuil to ask of him some nitrogen protoxide, with the aid of which I will be able, doubtless, to see carbon monoxide and oxygen flow.

"P. S.—I have just performed an experiment which gives my mind great peace. I have compressed some hydrogen to 300 atmospheres, and, after cooling to —28°, I have released it suddenly. There was no trace of mist in the tube. My gases (CO and O) are then on the point of liquefying, this mist not being produced except with the vapors near liquefaction. The (*prévisions*) prophecies of M. Berthelot are completely realized.

"LOUIS CAILLETET.

"December 2, 1877."

The control experiment with hydrogen, with its negative results, gives great conclusiveness to the experiments in which a positive result was obtained.

The letter had been deposited, sealed, with the Academy of Sciences at Paris on December 3, 1877.

He next turned his attention to nitrogen, compressed it to 200 atmospheres at 13° C. (55·4° F.), and on releasing it from pressure it condensed very perfectly, "like a pulverized liquid," giving "droplets of appreciable size," which gradually disappeared from the walls toward the center of the tube, forming finally a vertical column around the axis of the tube. The duration of the phenomena was about three seconds. On December 30, 1877, the experiment was repeated many times before several members of the Academy.

The next day he tried to liquefy hydrogen in presence of MM. Berthelot, Sainte-Claire Deville and

Mascart, obtaining evidences of the liquefaction of the gas, and repeating the experiment a great many times. He compressed it to 280 atmospheres, and, on sudden release, it formed an exceedingly fine and subtile mist which suddenly disappeared.

Air purified from carbon dioxide and from water produced the mist without difficulty.

Berthelot, in commenting on the liquefaction of hydrogen, says:

"The extreme tenuity of the liquefied particles which form this mist of hydrogen, a sort of disseminated glimmer (*lueur*), as well as their more rapid return to the gaseous state, are in perfect accord with the comparative properties of hydrogen and of the other gases." (*Comptes Rendus*, vol. lxxxv.)

The rival claims of Pictet and Cailletet are compared by Sainte-Claire Deville, who says that Cailletet's experiments were repeated in the École Normale on December 16, and succeeded perfectly. This was the day of his election as a correspondent of the French Academy of Science. The priority of discovery is awarded to Cailletet.

When we see later how much store Olszewski sets by his claim to have been the first to produce liquefied oxygen in quantity sufficient to be poured from one vessel to another, when we read between the lines of Cailletet's letter that he would have liked to produce a real visible bulk of liquid oxygen, we can appreciate Pictet's work at its full value, and feel that the two deserved at least equal honor.

The two worked quite independently and without knowledge of the scope of each other's work. It seems a pity that they could not have been associated as were Wroblewski and Olszewski five years later. It is the great chemist Dumas who, in the *Transactions* of the Academy of Science, calls attention to their ignorance of each other's work. It is pleasing to know that later in life they contracted an intimate friendship with each other.

Cailletet seemed to think that, as he had liquefied the constituents of air, the liquefaction of air itself was of little importance.

On trying air at 200 atmospheres, and on cooling the upper part of the tube with nitrous oxide, threads of liquid appeared on the walls of the tube. They were very agitated, and, on running down until they struck the mercury, they recoiled or drew back.

He felt that a control experiment was needed to determine if a liquid near its point of condensation would act in this way. Ether was selected on account of its high volatility. He poured it down a tube and found that it gave the same effect as he had seen in his compression apparatus.

Inspired by confidence from this control test, he increased the degree of compression in his apparatus until the mercury rose into the small tube within the refrigerating vessel, the pressure rising to 225 atmospheres and the liquid threads or streamlets increasing in number.

Continuing the compression until 310 atmospheres pressure was attained, the mercury reached the lèvel of the nitrous oxide, when it froze, stopping up the tube. The refrigerating apparatus was at once removed, when it was seen that the surface of the frozen mercury was covered with hoar frost, which he thought was solid air.

This closing the tube with a stopper (*bouchon*) of frozen mercury appears to him a method of very useful application in some of these investigations.

Cailletet showed much ingenuity in his methods, and the construction of his manometer for indicating high pressures and its standardization give a good sample of his work.

He first determined that glass yielded to pressure and returned perfectly to its original shape. He then constructed a manometer or pressure indicator by making what was practically a mercurial thermometer. The bulb was hermetically sealed in a steel reservoir full of water. On pressure being applied to the water, the bulb was squeezed and the mercury rose. The steel reservoir could be connected by a pipe to any fluid whose pressure was to be tested, as, for instance, to the water or mercury in his liquefaction apparatus. The manometer was kept at a uniform temperature by melting ice. The height to which the mercury rose gave the pressure.

The methods he adopted for testing its accuracy are striking. He fitted it with an index like a maximum thermometer and lowered it to known depths in the sea, in the harbor of Toulon, so that the water produced the known pressure for its calibration. He complains of the bad seas encountered. Another way was to lower it into an artesian well. In these cases he introduced maximum and minimum thermometers with it in order to secure corrections for temperature.

He also constructed an open end mercurial manometer, which was a long tube running up a cliff, and by maintaining mercury in it at different heights he produced a range of pressures from zero up to 34 atmospheres. This he used as a standard for testing the accuracy of his small manometers.

This was in the early period of his labors. It was not likely that such a practical and hard working scientific investigator could fail to see years later the chance which the Eiffel Tower, nearly a thousand feet high, offered for the construction of an open tube manometer. He interested M. Eiffel in the work. A soft steel tube was erected which ran up the framework of the tower. It was 4½ mm. (nearly $\frac{1}{6}$ inch) in internal diameter. Every 3 meters (nearly 10 feet) a projecting pipe with stopcock was placed, and to each of these a glass tube, in length slightly in excess of the 3 meters, was placed. Thus readings could be taken all the way up the tube. As each glass tube became filled, and the readings comprised within its length were completed, the stopcock was closed and more mercury was pumped in at the bottom.

The mercury came in from below. The steel tube dipped into a cistern, and a pump by hydraulic pressure forced the mercury into the cistern and up into the tube.

With this apparatus some 400 atmospheres of pressure could be reached.

Some rather curious corrections had to be applied. For a range of temperature of 30° C. (54° F.) the tower and steel tube expanded $\frac{1}{3000}$ of their length or height. This was a very minor matter. But the mercury for the same range expanded $\frac{1}{200}$. The heat expansion of the mercury, therefore, had to be corrected. The compressibility of the mercury and the diminished pressure of the air due to the great height were sufficient in extent to require correction also.

He tried manometers, as we have seen, by lowering them into watcr of great depths. The manometers operated by mercury rising in a glass tube. In the artesian well or in the harbor of Toulon the manometer was inaccessible, and an index was needed to show how far the mercury had risen in the tube.

This he secured by gilding the interior of the glass tube. As the mercury rose, it amalgamated with the gold and removed it from the glass. The portion of the glass tube stripped of gold showed how much of the tube had been filled with mercury. The arrangement operated like a maximum thermometer.

It cannot but impress the reader of the old time original papers on scientific work which have marked the steps of our progress that there is much good matter in them which has been forgotten. An original memoir ten years old is apt to be forgotten or to be treated as something which has been supplanted by more modern writings. But this view of the case is wrong and unjust, for the history and development of science is a most interesting study, and in these days, when the inductive method of teaching is so extensively employed, the old original papers by the great ones of the scientific world should receive much more attention than is generally awarded them. This book has been written from this standpoint. The bibliography of liquid air and liquefied gases testifies to the amount of material there is to be drawn upon.

Cailletet's work on his manometers shows a very good and conclusive method of measuring high pressures. His operations indicate an original cast of thought. After his great work on the liquefaction of oxygen by the use of his happily utilized pressure release he continued his work on gases. In 1880 he investigated the phenomena brought about by compressing a mixture of carbon dioxide and air. He found that the carbon dioxide was first liquefied and then disappeared as the pressure rose, which he interpreted as the solution of a liquid in a gas. It reminds us of the solution of a solid in a gas shown when a solution of a solid in a liquid is heated to a point above the critical one for the solution in question. Thus, if potassium iodide or chlorophyl is dissolved in alcohol, and the solution is heated in a sealed tube to 350° C. (662° F.), the whole disappears, and the solid is, so to say, dissolved or diffused in the gaseous alcohol. The observation is due to Hannay and Hogarth, page 23.

Cailletet noted the same thing with the liquid carbon dioxide and the gaseous air. He wished to have some test to determine when his carbon dioxide parted from the liquid state, and he sought a coloring agent for it. He thought that, if he colored it, the change from liquid to gaseous would be discernible. After some trials of different agents, he found a coloring matter which would dissolve in and color liquid carbon dioxide. It was the blue oil of galbanum.

Galbanum is a resin imported from the Levant, used in medicine, and of somewhat uncertain origin. Those who are interested in the archæology of science will find it mentioned in Exodus xxx. 34. The old name for it was chelbenah.

This ancient member of the pharmacopœia gave Cailletet the coloring matter he sought for. Liquid carbon dioxide dissolved it, and was colored blue thereby. On gasification the blue oil was deposited on the sides of the tube and on the surface of the mercury.

He investigated the peculiar striations which occur around the critical point, and concluded from the action of the coloring matter that they were liquid carbon dioxide. The disappearance of the meniscus was determined to be due not to liquefaction of the entire contents of the tube, but to gasification. The general phenomena presented by a mixture of carbon dioxide and air when highly compressed were studied, and the results are given in the *Comptes Rendus*, vol. xc.

Years later he returns to this question of coloring liquid carbon dioxide in order to determine the point of its gasification when heated in a sealed tube under pressure. He expresses some discontent with oil of galbanum and tries iodine. He sublimes this in his gas tube, so that portions of the glass collect a sublimate. He liquefies carbon dioxide in this tube, when it becomes colored by the iodine. On heating to disappearance of the meniscus, he finds that the gas or liquid in the lower part of the tube is blue, while that above is colorless, although iodine is there to color it.

A test with the spectroscope shows that the carbon dioxide colored with iodine gives the spectrum of iodine in solution, not of gaseous iodine. So the conclusion is reached that the disappearance of the meniscus is not necessarily synchronous with the attainment of the critical temperature.

To further examine the question, he tries an analogous experiment with two liquids, immiscible under ordinary conditions. Amylic alcohol and common alcohol, each with some water, lie one above another without mixing. He places the two in a sealed tube and applies heat. The line of separation between them begins to disappear, vanishes, and striations, such as seen with liquid and gaseous carbon dioxide heated to the critical temperature, appear.

He gives a new definition of the critical temperature, as follows: The temperature at which a liquid and a gas above it are capable of mutually dissolving each other in all proportions.

His condensing pump, without harmful clearance or lost space (*sans espace nuisible*), excited considerable attention. If a condensing pump has much clearance, if the piston or plunger does not go against the end of the cylinder as it expels the gas, as the pressure against which the pump works rises sufficiently high, no gas will be expelled, and the pump will do no work. This point is spoken of where Pictet's experiments are treated of, on page 165, and his way of getting over the difficulty, the coupling of two pumps, was spoken of.

Cailletet constructed a single acting plunger pump. It was placed with its cylinder vertical. The gas was forced out of its upper end. To avoid clearance he placed a quantity of mercury over the piston. As it rose, the mercury was forced into the clearance space, so as to completely fill it and thus suppress its injurious action.

It is evident that it is impossible to construct a pump without any clearance as they are ordinarily built. But the liquid piston obviates the trouble, and at each stroke every particle of gas is expelled, whatever may be the pressure against which it works.

Before Cailletet's pump was devised, Regnault had experimented with a mercury pump on somewhat the same principle. The Cailletet pump has, however, been accepted as a most valuable contribution to compressed gas work, and has been adopted by the Leyden University in its cryogenic laboratory. It has done much service in the hands of other investigators.

The cut gives a section of the barrel of the pump. *B B* is the barrel with plunger, *A*. The dark portion over the plunger is mercury. At *a*, *b* are packing rings of leather. *R* is the inlet valve, which is worked by a cam and lever system automatically. The neck, *O*, through which the gas enters, can be connected by a rubber tube to the source of supply. *S* is an ebonite valve through which the gas is forced by the plunger into the bonnet which surmounts the barrel. The tube, *T T*, delivers the compressed gas. A flexible copper pipe is connected to the tube, *T T*, and leads to the vessel in which the gas is to be condensed.

The operation of the pump is obvious. The plunger begins to rise, and the valve, *R*, closes. The plunger then drives the gas before it through the valve, *S*, and as it reaches the upper part of the cylinder, the mercury rising into the narrow tube below *S* expels the last traces of gas. As the plunger descends, an almost perfect vacuum forms above the mercury until the valve, *R*, is passed and opens, so that the gas to be compressed can enter. On the return stroke, this quantity is driven out through the valve, *S*.

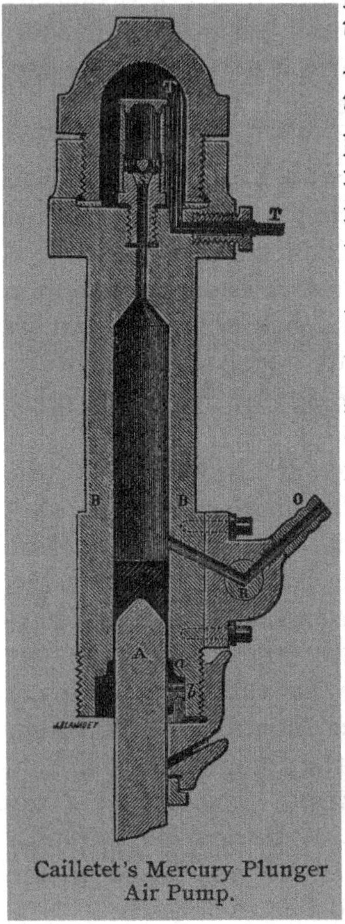

Cailletet's Mercury Plunger Air Pump.

If any mercury enters the bonnet or space above the valve, *S*, it cannot reach the gas reservoir, because the outlet tube, *T*, takes the gas from the top of the space. The pump was operated with a fly-wheel and crank motion, much like Natterer's pump. Sometimes a screw valve was placed on the summit of the bonnet to make possible the expulsion of all air from the pump.

The base of the barrel screwed into a socket or base piece, which held some glycerine or mercury, to insure the tightness of the packings, *a* and *b*.

The presence of mercury made the lubrication

problem somewhat troublesome, as ordinary oils and grease coming in contact with the mercury formed almost solid compounds. Eventually Cailletet adopted vaseline and glycerine as lubricants.

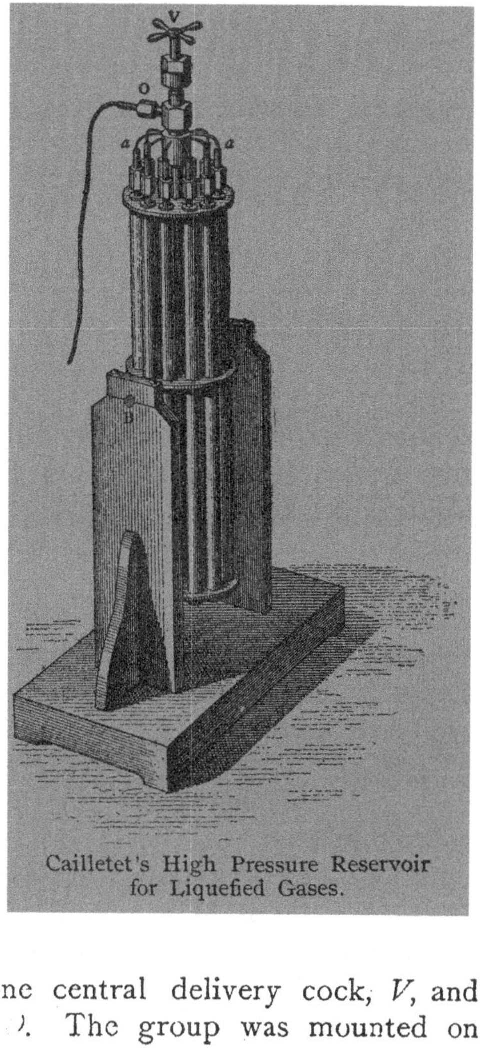

Cailletet's High Pressure Reservoir for Liquefied Gases.

With this pump a man, without over-exertion, readily liquefied 400 to 500 grammes of carbon dioxide in an hour.

Recognizing the danger of the larger cylinders used for holding liquefied gases, Cailletet replaced them by a group of nine copper tubes, arranged in a cylindrical group, and all connected by small copper tubes, *a a*, to one central delivery cock, *V*, and outlet coupling). The group was mounted on trunnions, *B*, in a frame, as shown in the cut. The tubes had a capacity of about four liters.

The mercury pump without lost space (*sans espace nuisible*), as invented by Cailletet after Regnault had experimented with one, is of special interest, and has been very often used in liquefied gas investigations. One of the most celebrated high pressure gas laboratories, that of the University of Leyden, uses it in a modified construction. The mercury no longer lies on the plunger, but is beneath its end. A U-shaped tube constitutes the pump barrel. In one limb the plunger works downward. The bend of the tube is filled with mercury, and the outlet for gas as compressed is at the top of the other limb.

All through the history of investigations on this subject we find at intervals Cailletet's pump mentioned; so it has survived a long time as things go in this age of progress. The new demand is for a pump that will continuously and powerfully compress a gas. Formerly it was a single sample of gas at a time which was to be compressed. This was effected by a screw or other device, as explained and described in many places in this book. But when Pictet, in 1877, established his double cycle liquefaction of gas, he instituted a method calling for a pump with constant delivery at high pressure, and his method has been utilized in some shape or form by most subsequent investigators until within the last few years. It is by no means abandoned yet, and Cailletet's pump is still, in improved form, doing service at the cryogenic laboratory of the University of Leyden.

Cailletet and Hauteville, in 1882, approached the difficult task of determining the specific gravity of liquid oxygen in the following way:

One volume of oxygen was mixed with seven volumes of carbon dioxide. The mixture was submitted to pressure while maintained at a temperature exceeding the critical temperature of the relatively easily liquefied carbon dioxide. Then, after the compression was effected, the temperature of the mixture was lowered and the two gases liquefied together without separation. Numerous experiments with other gases had shown that there was no reason to expect any shrinking, except in very slight degree, upon mixing two such liquids. The specific gravity of liquid carbon dioxide was easily determinable and was accurately known. The mixture of liquid oxygen and carbon dioxide was perfectly manageable, and its specific gravity was determined with ease, and by simple alligation the following results were obtained. At the melting point of ice, 0° C. (32° F.) and at —23° C. (—9·4° F.), it was for various pressures expressed in atmospheres:

Pressure.	0° C. (32° F.)	—23° C. (—9·4° F.)
200	0·58 sp. gr.	0·84 sp. gr.
275	0·65 "	0·88 "
300	0·70 "	0·89 "

As a control, a similar experiment with nitrous oxide, substituted for carbon dioxide, gave at 300 atmospheres and at —23° C. (—9·4° F.) a specific gravity of 0·94.

Another of Cailletet's classic discoveries is the use of liquid ethylene as a cooling agent. According to him, it liquefies at the following pressures and temperatures:

45	atmospheres at	1° C. (33·8° F.)
50	" "	4° C. (39·2° F.)
56	" "	8° C. (46·4° F.)
60	" "	10° C. (50° F.)

Its critical temperature is about 13° C. (55·4° F.), while that of its predecessor in the refrigerating line, carbon dioxide, is 31° C. (87·8° F.) Using a carbon bisulphide thermometer, he reached a temperature of —105° C. (—157° F.) in liquid ethylene, while in liquid nitrous oxide he only reached —88° C. (—126·4° F.)

He made the ethylene by the old method of heating together alcohol and concentrated sulphuric acid. The latter, with its high affinity for water, takes the elements of water from the alcohol, and gaseous ethylene is evolved. This gas he liquefied by the use of his mercurial pump just described. He found it far from manageable by his appliances, and first employed it as a refrigerant in the form of a jet, reminding us of Thilorier's *chalumeau de froid*, or cold jet blowpipe, spoken of in a preceding part of this book (page 141).

In its release from confinement it goes into the gaseous state, not solidifying into snow, like carbon dioxide.

The classic interest of this discovery lies in the great use that subsequent investigators have made of liquid ethylene as a refrigerant. Notably is this the case with the work done in the Royal Institution by Dewar. One of the most striking features of his work was the number of cylinders of liquid ethylene which he prepared for his liquefactions. Such was the comment made by Prof. George Barker on his visit to the Royal Institution. The quantity of liquid ethylene was as remarkable in its way as the liquefaction of air itself, and the manufacture of this ethylene was one of the principal sources of expense incurred in the Dewar liquefactions.

The ease of liquefaction of ethylene, its reasonably high critical temperature and the high degree of cold produced by its evaporation, make it a particularly valuable and manageable agent. The difficulties Cailletet experienced have disappeared with the improved appliances of fifteen years later.

Ethylene is a very old acquaintance and a compound that, in giving its luminosity to coal gas, has played an important role in technology.

An objection to ethylene, as a refrigerating agent, is its cost. It is no great matter to make a few cubic feet of the gas from alcohol and sulphuric acid; but when it comes to condensing the gas to a liquid with many hundredfold reduction of volume, the cost becomes very great. A 5 or 10 gallon cylinder of the liquid represents immense expenditure of alcohol. Cailletet's very inartificial way of using ethylene as a cold jet blowpipe and letting the gas go completely to waste complicated the difficult gas liquefactions by the introduction of a very serious factor of expense.

In a subsequent paper we see that he appreciated this state of affairs and tried to work with less waste and to introduce a rational economy into his process.

In 1883 Cailletet speaks of a continuous liquefying apparatus, but declines to describe it. Hitherto he had operated with small quantities of liquid ethylene at a time, by the use of his mercury condensing pump, and had applied the ethylene as a jet, but now he uses a closed cycle. The ethylene circulates through a steel cylinder, being released from compression as it enters, so as to take the gaseous form, and reducing the temperature greatly on the latent heat principle. Through the steel cylinder a tube passes, so that the two represent a condenser of the type of the well-known Liebig's condenser, similar to Pictet's apparatus. The pump draws out the gas from the cylinder and compresses it to the liquid state, so that it is ready to expand again as it enters the cylinder. He got an almost complete vacuum in the cylinder of his condenser, and a very low temperature resulted.

The arrangement is practically that of Pictet of 1877. Cailletet in 1883, and Dewar in 1890, and at later periods, bear testimony to the good quality of Pictet's early work in the arrangement of apparatus they adopted, and which was based on Pictet's apparatus, illustrated in this book.

Cailletet hoped to get oxygen in large quantities by the use of this new apparatus, evidently appreciating the defect inherent in the Colladon apparatus, which quite excluded the idea of operating on large quantities of gases, and which produced them in a necessarily non-continuous process. It will be remembered that it was the Colladon apparatus which Cailletet had adopted in his work of 1877.

A very ingenious method of producing low temperatures was studied by Cailletet, and his paper on the subject was published in 1885. He effected the evaporation of a liquefied gas with accompanying reduction of temperature by passing a stream of a cold gas through it. He placed a tube with ethylene within a vessel of dry air, and by blowing cooled and dry air or hydrogen through it accelerated its evaporation until its temperature fell to —136° C. (—212·8° F.) By such a process he produced cold sufficient to liquefy oxygen, the latter being compressed to the requisite extent.

Working with M. Bonty, in the same year, he made quite an elaborate series of experiments on the electrical conductivity at low temperatures of a number of metals—copper, mercury, silver, aluminum, tin, magnesium, iron and platinum. He suggests the availability of copper wire as a means for determining low temperatures, by its decrease of electrical resistance as the temperature falls. This suggestion is interesting, in view of subsequent developments. A passage from the paper on the subject will be of interest:

"It seemed probable that this resistance would become extremely small, and consequently the conductivity very great, at temperatures below —200°, although our first experiments did not permit us to form a definite idea of that which would occur under such conditions." (*Comptes Rendus*, vol. c., page 1189.)

This is in strict accord with the facts as ascertained by other experimenters at later periods.

In 1888 we have from him a comparison of five methods of determining low temperatures. They were the following: 1. A thermo-couple of iron and copper. 2. A platinum wire resistance. 3. A thermo-couple of pure platinum and of an alloy of platinum and rhodium. 4. An ingot of platinum used in conjunction with a calorimeter. 5. The hydrogen thermometer. Boiling water, melting ice and boiling methyl chloride, at atmospheric pressure, supplied the three fixed points for his scale, and he obtained very closely according figures with all of these methods. The temperatures of boiling ethylene and of boiling nitrous oxide were determined as tests of accordance of results.

Cagniard de la Tour had long ago tried to determine the point at which the meniscus of water disappeared when it was heated in a sealed tube. Pure water attacked the glass tube so actively that he could not produce the disappearance. Cailletet took up the question and tried the experiment in a metal tube, with pure water, applying a mathematical calculation to determine the desired point. The older observer had added chemicals to the water to diminish their action on the glass. Cailletet discerned in the presence of the chemicals a source of error and recognized the importance of performing the experiment with pure water. The description will be found in the *Comptes Rendus*, vol. cxii.

CHAPTER X.

SIGMUND VON WROBLEWSKI AND KARL OLSZEWSKI.

Wròblewski's life—Banishment from his native country—Early scientific work—His association with Olszewski—Study of Cailletet's methods—Their apparatus—Defective position of the hydrogen thermometer—Liquefactions of oxygen, carbon monoxide and nitrogen—Ethylene data—Solidification of carbon bisulphide and alcohol—Determination of the critical pressure and temperature of oxygen—Liquefaction of hydrogen—Use of a thermoelectric thermometer—Electric resistance of metals at low temperatures—Two liquids from air—Olszewski's individual work—Apparatus for producing liquid oxygen in quantity—Comparison of platinum resistance and of hydrogen thermometers—Determination of hydrogen constants.

As a serious investigator in the realm of the liquefaction of gases, no one can be cited who surpassed the Polish scientist Sigmund von Wroblewski (pronounced Vroblevski). He was born in Grodno, Poland, in 1845 Grodno is a province which went to Russia in the partition of Poland and figures in the final partition of 1815 as part of Russia. The kingdom of Poland, as arranged by the Congress of Vienna at the same time, remained as a separate kingdom and intact, although its monarch was the Czar of Russia. Then there was a long series of political disturbances and bloodshed, culminating in the disturbances of 1861–64, and Russia succeeded by the most arbitrary enactments and severe measures in suppressing the insurrections and in assimilating the so-called kingdom of Poland.

Wroblewski took part in the uprising as a Polish patriot, and was sent to Siberia in 1863, where he spent four years. His friends had influence, and managed to obtain his release from exile, to the extent of being allowed to live in an obscure Russian town. Eventually he was released from surveillance and went to Germany, visiting Heidelberg and Bonn, meeting Kirchoff and Clausius. He had a cosmical theory which was not received by either the physicists of Heidelberg or of Bonn with any encouragement. At the University of Berlin he consulted Prof. Helmholtz, who started him to work on physical investigation touching his new theory, and he completed two years of work under the many-sided and brilliant German. He published papers bearing on gases which received the honor of attracting the attention of Clerk Maxwell. His principal work on high pressure and low temperature applied to gases dates from his knowledge of the work of Cailletet on the same subject. He spent some time at the École Normale, in Paris, and saw and studied Cailletet's work. He had as associate Karl Olszewski (pronounced Olshevski), in the writing of the initial of whose Christian name a certain amount of confusion obtains, as it is sometimes written K, for Karl, and sometimes C, for Charles. The association between the two in their early work of 1883, and thereabout, is very intimate. In *Wiedemann's Annalen*, 1883, is published an article which gives the full account of their first important work in the liquefaction of gases. The authorship is given a dual form. The title reads in translation, "On the liquefaction of oxygen, nitrogen, and carbon monoxide, by Sigmund v. Wroblewski and Karl Olszewski." The article, it is impossible to believe, was written by anyone but Wroblewski, but when in its course anything is to be attributed to a single investigator, the expression "*einer von uns*" ("one of us") is always used.

Wroblewski died in 1888. As early as 1884 he predicted that liquid air would be the refrigerant of the future. His emotions, had he lived to see what has been done in the liquefaction of air, can only be imagined. The principal reason for his belief in the capabilities of liquid air was that it did not have to be prepared like carbon dioxide, sulphur dioxide, ethyl chloride, or ethylene, that the atmosphere gave an inexhaustible supply of matter adapted for the function of refrigeration and for use in a cooling cycle.

Wroblewski, in the early days of the liquefaction of gases, in 1885, pointed out the method of the future. In the light of what has been since then accomplished, a translation of his remarks from the *Wiener Sitzungsberichte* reads almost like a prophecy:

"The essential step forward which should be made with regard to the extension of the method is to change it so that we may be prepared to pour oxygen as we pour ethylene to-day. It is my conviction that the thing will only be successfully carried out when we return to Pictet's method, and by cycles of various liquefied gases make a cascade of temperatures whose last step will produce the stream of liquefied oxygen."

It is precisely by carrying out such a line of work that Dewar won fame for himself and the Royal Institution.

The carefully prepared article in *Wiedemann's Annalen* is an example in its way of how a scientific paper should be written. There is in its aspect and tenor such sincerity and so careful an avoidance of anything like self-assertion that it is at once convincing and impressive.

These investigators were subsequently attached to the University of Cracow, and much of their work dates from that city. The results are published in various languages. There is no need to study Polish to read them.

"One of us," Wroblewski, while in Paris studied Cailletet's apparatus and methods, and had an apparatus made by a Paris mechanic, E. Ducretet, for the prosecution of researches on liquefied gases. The point is made that it is superior to the Cailletet apparatus of that early date because it could be used with five or six times as much gas as could be used in Cailletet's apparatus.

The apparatus may be considered in two divisions—one the condensing apparatus by which the gas to be experimented on was subjected to pressure, the other the refrigerating apparatus for cooling it below the critical temperature.

We reproduce the cuts of the apparatus from *Wiedemann's Annalen.* It will be seen that the gas compression apparatus is practically a copy of Cailletet's apparatus, so that the apparatus goes back to the days of Colladon. In the gas refrigerating portion will be found a reminder of Pictet's circuits, not as yet fully utilized by the Polish scientists.

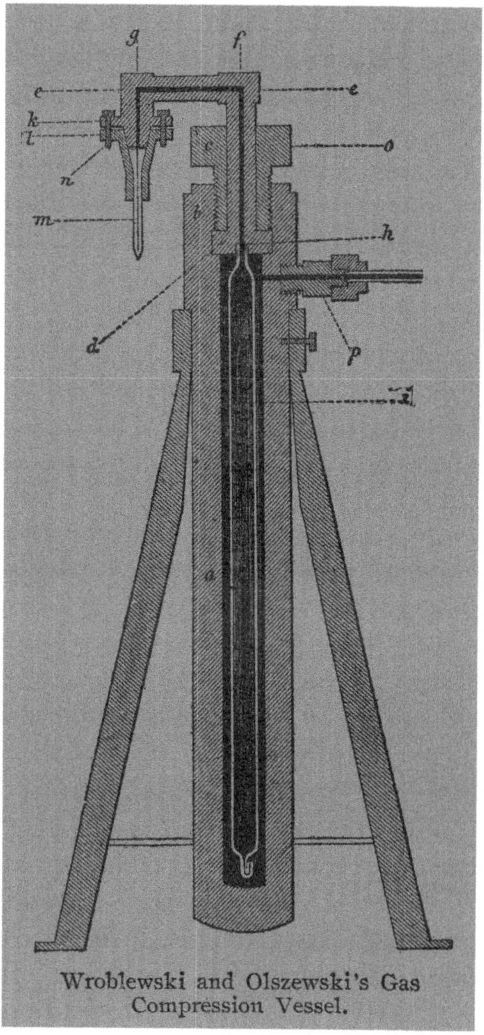

Wroblewski and Olszewski's Gas Compression Vessel.

The gas tube, *i*, is designed to hold about 200 cubic centimeters of gas. It has an upturned capillary tube at its bottom. A very thick-walled capillary tube extends from its top and bends downward. The cylinder, *a*, which contains the gas tube, is of heavy cast iron. Very exact dimensions are given in the paper in *Wiedemann's Annalen* already cited. The general dimensions are stated as 58 centimeters (23·2 inches) deep and 8·5 centimeters (3·4 inches) wide. *c* and *o* is a bronze piece which forms a tight connection between the gas tube, *i*, and the upper tube, *c*, *f*, *g*, *e*. A very strong steel tube runs through the orifice in the piece, *d*. To get it in place the horizontal portion of the piece in question was sawed through horizontally in the line, *e*, *c*, and bored downward from *g* and *f*. The steel tube was inserted in place, a groove along the line, *e*, *e*, receiving it. The piece which was sawed off was replaced and brazed in its former place, so as to surround the steel tube.

At the end, *h*, the steel tube expands, and the glass gas tube, *i*, is cemented into it. At *k* the bronze steel-lined piece has a conical end. *m* is a glass tube cemented in place, and all is secured by a coned piece, *l*, with screws, *n*, as shown, the screws uniting all parts to an airtight joint.

At *p* a tube is connected which leads to a force pump.

The next illustration shows how the apparatus was set up for the liquefying of gases in the downwardly extending tube from the compressing apparatus. This cut is also an exact reproduction of the cut given in *Wiedemann's Annalen.*

We have, as before, the vessel, *i*, with its steel containing vessel, *b*, only the top of which is shown. The capillary tube, *q*, was 0·9 centimeter (0·36 inch) external diameter and a little over 0·2 centimeter (0·08 inch) internal diameter. The glass vessel, *i*, was filled with the gas to be experimented with.

A jar, *y*, has calcium chloride at its bottom to keep the air within it perfectly dry. A second vessel, *s*, is set into it airtight with an india rubber stopper. The vessel, *s*, is provided with an india rubber stopper of its own, perforated for three tubes. One is

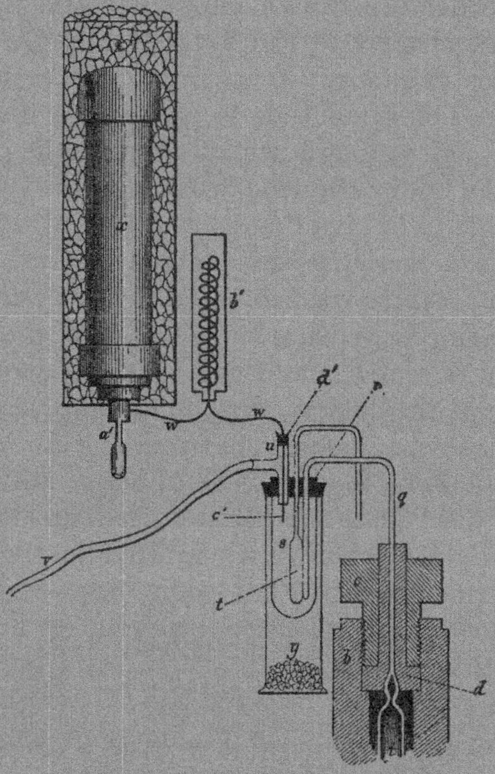

Wroblewski and Olszewski's Apparatus for Liquefying Gases.

the end, *q*, of the gas tube, *i*, the other the stem of the hydrogen thermometer, *t*. The third receives a T shaped tube, *u*. Liquid ethylene is contained in the cylinder, *x*, where it is kept cool with ice and salt. The liquid ethylene is withdrawn at *a*, through a thin tube, *w*. This tube is coiled into a cooler, *b′*, charged with liquid and solid carbon dioxide. This brings it down to a very low temperature.

As needed it is drawn into the vessel, *s*. An air pump connected to the T tube, *u*, by the tube, *v*, produces an almost full vacuum in the vessel, *s*. The upper end of the T tube is provided with an india rubber cork through which the tube, *w*, passes air-tight, the liquid ethylene dropping from *c*.

The gas to be experimented on was introduced into the tube, *i*, mercury was contained in the vessel, *b*, and the pressure was increased to any desired extent by pumping water into *b*. The end of the gas tube, which was sealed and bent down, was cooled by admission of the cooled ethylene into the vessel, *s*, and this vessel was pumped out by an air pump, so that it was kept down to a pressure of but 2½ millimeters of mercury, which is a small fraction of an atmosphere. The ethylene, when first admitted to the vessel, *s*, boiled tumultuously, but soon quieted down and kept slowly boiling, thereby producing a very low temperature.

Each experiment required 200 to 300 grammes of ethylene and about 400 grammes of solid carbon dioxide. Very little ethylene was lost.

The apparatus worked well. The only trouble chronicled was due to the mercury freezing in the capillary tube, which brought about an explosion which did no great injury.

The temperatures were taken by the hydrogen thermometer, *t*, whose bulb, it will be observed, is placed in the refrigerating vessel, not in the gas experimented with. Thus the temperature recorded is that of the environment of the sample, not that of the sample itself, which is a defect worthy of comment.

While on the subject of thermometers, it may be noted that there occurs in the *Wiedemann's Annalen* article an interesting statement to the effect that Natterer told "one of us," orally, that he filled his low temperature thermometer with phosphorus chloride. This gives us a glance at the work of a preceding generation and is mentioned elsewhere in this book.

The results obtained with this apparatus were very good. Oxygen liquefied at —130°. C. (—202° F.) and at a pressure of a little over 20 atmospheres. It was a colorless fluid, the slight blue tint not showing, presumably because of its slight amount. It had a flatter meniscus than that of carbon dioxide. On reducing the pressure to a relatively small degree it foamed, evaporated from the surface, and on further reduction, boiled throughout its entire mass.

The work of these investigators at about this period is the subject of other papers in the *Comptes Rendus* and elsewhere.

In the *Comptes Rendus*, vol. xcvi., is given the dispatch announcing Wroblewski's liquefaction of oxygen. It was received by M. Debray, secretary of the Academy of Sciences, on April 9, 1883, from Cracow. It reads as follows:

"Oxygène liquéfié, completement liquide, incolore comme l'acide carbonique. Vous recevrez une note dans quelques jours."

"Oxygen liquefied, completely liquid, colorless like carbonic acid. You will receive a note in a few days."

The "note" which follows is given in the same volume of the *Comptes Rendus* and alludes to Cailletet's ethylene paper (ibid., vol. xciv., page 1224). The authors say that Cailletet did not fully satisfy himself. Wroblewski and Olszewski, with apparatus made by "one of us" ("*un de nous*"), who was in this case Wroblewski, and using a quantity of oxygen, effected the liquefaction. They found liquid oxygen colorless and transparent, very mobile, and giving a sharp meniscus.

With boiling ethylene in approximate vacuum they got a temperature of —136° C. (—212·8° F.) by the hydrogen thermometer. They found that at the atmospheric pressure ethylene boils at —102° to —103° C. (—151·6° to 153·4° F.), and not at —105° C. (—157° F.) The following data for oxygen were determined on April 9:

At temperature of —131·6° C. (—204·9° F.) begins to liquefy at 25·5 atmospheres.

At temperature of —133·4° C. (—208·1° F.) begins to liquefy at 24·8 atmospheres.

At temperature of —135·8° C. (—212·4° F.) begins to liquefy at 22·5 atmospheres.

They took advantage of their ethylene apparatus to try some other experiments in the direction of freezing carbon bisulphide and alcohol.

Carbon bisulphide froze at about —116° C. (—176·8° F.), alcohol became thick like sirup at about —129° C. (—200·2° F.), and froze a degree lower, —130° C. (—202° F.)

On April 16, 1883, another dispatch was received by the secretary of the Academy of Sciences, telling of the same investigators' liquefaction of nitrogen:

"Azote refroidi, liquéfiée par detente. Menisque visible, liquide incolore."

"Nitrogen cooled, liquefied by release. Visible meniscus, colorless liquid."

The note which gives the details of the liquefaction of nitrogen says that they exposed nitrogen at —136° C. (—212·8° F.) to a pressure of 150 atmospheres. On sudden release there was a tumultuous ebullition ("*aufbrausen*") like that of carbon dioxide in a Natterer's glass tube of carbon dioxide (page 23) when it is plunged into water which is a little warmer than the critical temperature of carbon dioxide. Then they tried a partial release from pressure, lowering it to 50 atmospheres, when the nitrogen liquefied completely with a meniscus. It remained a few seconds only. It was colorless and transparent.

On April 21, 1883, the following dispatch was received by the Academy from the same investigators:

"Oxyde de carbone liquéfié dans les mêmes conditions que l'azote. Ménisque visible. Liquide incolore."

"Carbon monoxide liquefied under the same conditions as nitrogen. Meniscus visible. Colorless liquid."

Hydrogen they failed to liquefy. It was cooled to —136° C. (—212·8 °F.), compressed to 150 atmospheres, then was suddenly released, but not even a mist appeared. Boiling oxygen is recommended as a cooling agent, but the impetuousness with which it boiled was a great obstacle to its use. Even at one atmosphere of pressure it proved uncontrollable. The duration of its ebullition was very short, and this proved an objection. Eight years later, in 1891, Olszewski overcame this trouble by bubbling hydrogen through it gradually. Cailletet's production of cold by bubbling a gas through a volatile liquid, as described on page 201, may be noted also. By a thermo-electric couple its temperature was determined. It is given as —186° C. (—302·8° F.)

Nitrogen was compressed and cooled with boiling oxygen without result, but on sudden release from pressure it formed snow-like crystals of remarkable size.

In 1883 Wroblewski and Olszewski attacked the problem of determining the specific gravity of pure oxygen. They introduced a known quantity of oxygen into their apparatus and liquefied it as completely as possible. This gave them an approximation, if they neglected to take into account the unliquefied gas which lay above the liquid. To determine what value this unliquefied portion had, a control experiment was done with liquid carbon dioxide whose specific gravity was known, the experimenters using Andréeff's determination (*Liebig's Annalen*, vol. cx., page 1). The calculations are too complicated to be here reproduced. The result obtained for oxygen at about —130° C. (—202° F.) and the pressure of liquefaction was 0·899.

Wroblewski, still longing to produce liquid oxygen in quantity, says, in December, 1883, that it is merely a question of appliances to produce liquid oxygen, but acknowledges that he has never succeeded in producing oxygen in the condition of a static liquid. Any attempt to use the refrigerating effect of oxygen, he said, involves its use at the instant of production or cessation of pressure. Such danger of explosion attended attempts in this direction that masks were worn.

A valuable suggestion would seem to be the one made in 1884, when Wroblewski suggests the use of liquid marsh gas as a refrigerant. In its properties it is adapted to fill the gap which exists between liquid ethylene and liquid oxygen. The honor of being the first in the field with this suggestion was afterward claimed by Cailletet. Dewar, however, was able to show that he had suggested the use of liquid marsh gas as far back as 1883, which antedates Wroblewski, and Cailletet's date goes back to 1881.

After this period the two scientists appear as individual workers. The path started on the lines of Cailletet's and Pictet's work led to direct experimental determinations, but these appear in later work. The early apparatus, just described, did not lend itself to thoroughly reliable temperature observations. Indirect methods of dealing with problems had to be used, and in some cases data were reached on almost purely theoretical grounds. This was done to some extent quite recently, and the hydrogen data were determined with fair approximation partly from a theoretical basis.

Much ingenuity appears in the methods of attacking the problems which presented themselves in the course of their experimentation. As an example may be cited the determination of the critical temperature and pressure of oxygen (*Comptes Rendus*, vol. xcvii.)

Oxygen gas was liquefied in the downwardly bent tube, *q*, of the apparatus, page 209, by the aid of boiling ethylene contained in the vessel, *s*, as already described. As the oxygen liquefied its level rose in the tube, *q*, and eventually reached a point above the level of the liquid ethylene in *s*. Now it is evident that, as the liquid oxygen reaches a point in the gas tube above the ethylene, the temperature of its upper layers is higher, and the more it rises, the higher is this temperature. As the temperature increases, the pressure necessarily rises.

At last a point is reached when evidences of the critical state begin to show themselves. The meniscus flattens, the line of demarkation between liquid and gas becomes indistinct and at last entirely disappears. The only way to trace the position of any separating level is by the difference of refractive power of the different layers in the tube. The description as given by Wroblewski exactly describes the phenomena observed in a Natterer's tube (page 23).

If the pressure is lowered, the temperature of the oxygen falls, liquefaction ensues, and the meniscus again forms. Working in conjunction with Olszewski, the investigator found that this phenomenon of the critical state occurred always at about the pressure of 50 atmospheres.

The pressure of oxygen under these conditions is so high and its temperature so low that it appeared desirable to exercise some sort of a check upon this experiment. The same tube was charged with liquid carbon dioxide overlaid by the gas, in exact analogue with the conditions of the oxygen experiment The boiling ethylene was replaced by melting ice, and warm water at 50° C. (122° F.) surrounded the upper part of the tube. Hence, within the length of the gas tube the temperature had a range of 50° C.

Pressure was applied, and at 35 atmospheres traces of liquid carbon dioxide appeared in the bottom of the tube, which was the cold part. The gas kept on liquefying until the liquid rose above the level of the melting ice and began to reach the warm portion of the gas tube. The pressure increased as the liquefied carbon dioxide attained in its upper layers a higher temperature.

As the pressure approached 76 atmospheres the meniscus became flat, then indistinct, and eventually disappeared. The critical state was reached. On lowering the pressure, the liquid diminished in amount, the level fell, and the upper layer reached a cooler part of the tube. The meniscus at once showed itself again. The appearance and disappearance of the meniscus evidently took place at a point of the tube where the critical temperature existed. The pressure in the apparatus when the phenomena described took place was the critical pressure.

The attempt was made now to ascertain the critical temperature of oxygen—a far more difficult factor to determine. A small quantity of oxygen was liquefied in the apparatus, so that it was below the level of the liquid ethylene. The latter was boiling under exhaustion so as to give a very low degree of temperature. The exhaustion was stopped and the temperature of the ethylene began to rise. The meniscus was watched.

Two things were occurring in the tube. The temperature was rising and the pressure increasing as the ethylene became warmer. Sooner or later the balancing point, the critical state, would be reached and the disappearance of the meniscus gave the indication. This was watched for, the temperature of the ethylene being constantly observed.

The observations were extremely difficult, and Wroblewski gives the figure of —113° C. (—171·4° F.) in his own words, "as the first approximation to the critical temperature of oxygen." The temperature we now know was too high by nearly 6° C.

Cailletet had brought before the French Academy of Sciences his liquefaction of hydrogen (page 184). He had on release from pressure obtained a mist or fog, which he claimed was due to liquid hydrogen. Naturally some doubt was felt about it.

Wroblewski had tried it, and in an early number of the *Comptes Rendus*—early as regards its date—referring to the history of the liquefaction of oxygen and of the "permanent gases," says that he tried Cailletet's experiment and failed.

On January 4, 1884, the following dispatch from Wroblewski was received by the French Academy of Sciences:

"Hydrogène refroidi par oxygène bouillant s'est liquefié par detente."

"Hydrogen cooled by boiling oxygen has been liquefied by release."

Debray commented on the dispatch and says that this experiment confirms Cailletet's experiment.

In the *Comptes Rendus* of February, 1884, Wroblewski tells of his liquefaction of hydrogen. He compressed hydrogen to 100 atmospheres in a glass tube whose general dimensions were from 0·2 cm. to 0·4 cm. (0·08 inch to 0·16 inch) in internal diameter and 2 cm. (0·8 inch) external diameter. It was arranged for very sudden release of pressure. The tube was surrounded with boiling oxygen in order to reduce the temperature of the hydrogen. On sudden release of pressure the hydrogen gave the mist as in Cailletet's experiment of 1882.

To determine the temperature a thermocouple was used, which was connected to a galvanometer which could show a potential difference of $\frac{1}{400000}$ volt, which corresponded to half a degree on the thermometric scale. It was standardized by comparison with a hydrogen thermometer.

It was known that the electric resistance of metals falls with the reduction of temperature. As early as 1885 Wroblewski had tried silk-covered copper wire, cooled to a temperature of —200° C. (—328° F.), and found that its resistance was less than one-hundredth of what it was at the temperature of boiling water. He says that oxygen and nitrogen, in the liquid state, are among the most perfect insulators known. He says that the electric resistance of copper, at a temperature approaching that of boiling nitrogen, tends to become zero—the conductivity approaches perfection.

This view has been very prominently brought forward again by Dewar and others, and Elihu Thomson goes so far as to believe that in liquid gases a useful reducer of electric resistance for power distribution may be found. It is certainly very captivating to think of a thin copper wire in a pipe filled with liquid air carrying the energy of Niagara Falls over hundreds of miles of country.

An experiment which excited much comment, and which now, in these days of wholesale liquefaction of air, is almost lost sight of, was described by Wroblewski, who, in 1885, in liquefying air, produced from it two liquids superimposed and which remained separate for some minutes. He managed to withdraw, by a metallic tube, samples from each layer for analysis—rather a delicate operation, it would seem. On analysis, the lower layer, after gasification, gave a little over one-fifth of its volume of oxygen (21·28 per cent. to 21·5 per cent. oxygen). The upper liquid gave a little over seventeen-hundredths of its volume of oxygen after gasification (17·3 per cent. of nitrogen).

Wroblewski had used various thermometers for determining the low temperatures which he obtained in his experiments, the hydrogen-filled thermometer seeming eventually to give him most satisfaction. Cailletet had used various thermometers, finally tending to the hydrogen one. Pictet had adopted a very indirect method of calculating temperatures, and the thermo-couple had also been employed, as we have just seen.

In 1885 Wroblewski published a paper embodying his experiments on the relations existing between temperatures as determined by the hydrogen thermometer and a thermo-electric couple of copper and German silver.

After this year but little appears under the name of

this distinguished investigator. He seemed to possess the rare faculty of not disputing with any of his confrères. The disputes as to priority in the liquefaction of gases are very numerous and extend over the greater part of a century. Wroblewski was fortunate in not being involved in any of them, as far as his own statements are concerned at least.

Wroblewski and Olszewski worked together for a number of years, but the latter scientist continued the same line of work alone up to a recent period. In the *Philosophical Magazine*, March, 1895, he published a *résumé* of his work, incidentally giving vent to a certain amount of feeling and attacking Dewar and Pictet.

In 1885 Olszewski made what may be called an approximate liquefaction of hydrogen. He mixed two volumes of hydrogen with one volume of oxygen and liquefied the mixture successfully. The mixture was colorless. On release from pressure it lost most of its hydrogen. The residual liquid lasted for some time at the atmospheric pressure.

He is much interested in showing that he produced oxygen in quantity large enough to pour from one vessel into another. In October, 1890, he produced 100 cubic centimeters before an audience, and in July of the succeeding year, also before an audience, he produced 200 cubic centimeters. He lays great stress on this achievement.

His apparatus, by which he produced oxygen in what were large quantities for the period, was very simple. Its essential feature was the use of a steel cylinder of small capacity in which the oxygen was liquefied. This took the place of the glass tube in which the gases were liquefied in the original Wroblewski and Olszewski experiments.

In 1883 and the subsequent years the two associated investigators had liquefied gases in glass tubes. The almost capillary tube of their early experiments was changed sometimes for a larger one. Thus the following are given as the dimensions of a tube in which many liquefactions were carried out: The tube was 30 centimeters (about 12 inches) long and 14 to 18 millimeters (0·56 to 0·72 inch) in internal diameter. The walls were 3 to 4 millimeters (0·12 to 0·16 inch) thick.

All the "permanent" gases then known, from which argon, helium and the companions of argon must be excluded, for they were not yet discovered, had been liquefied in this apparatus, as already described, and nitrogen, carbonic oxide, nitric oxide and marsh gas had been solidified.

It will be observed, especially if the cut of the 1883 apparatus (page 209) be inspected, that no means were provided for drawing off the small amount of liquefied gas which might be produced in the glass tube. If an attempt had been made to substitute a large glass bulb for the tube, it would never have stood the strains due to changes of temperature and high pressure. By the repetition of numberless liquefactions, the conditions necessary to produce them became so accurately known that it was no longer necessary to see the liquefaction to know that it was produced. The necessity for a transparent vessel had ceased.

Olszewski accordingly substituted for the glass

tube a small steel reservoir. This would stand the pressure without danger of explosion, and was so good a conductor of heat that the most sudden changes of temperature had not the least effect upon it in the direction of causing it to break.

This apparatus was described in 1890 in the *Bulletin internationale de l'Academie de Cracovie.* While Olszewski, in the *Philosophical Magazine* article, seems to indicate that his work has not been fully enough appreciated, he makes very evident one reason. He gives the list of his original papers. So many of them appeared in the *Cracow Bulletin*, whose title is given above, that they were deprived of the circulation which was their due and which would have been secured by a wider publication in the German, French and English scientific annals.

But Olszewski's steel reservoir, like Pictet's liquefaction tube, was provided with a cock by which its contents could be withdrawn, and this certainly was an advance over a sealed glass tube. The probabilities are that in 1883 the possibility of handling liquid gases at atmospheric pressure like so much water was undreamed of.

The mechanically bad feature of Pictet's old apparatus was present in this one, which comes some thirteen years later. The liquid was drawn from a reservoir in which it was confined under enormous pressure. The outrush of the almost uncontrollable fluid must have given some trouble to the experimenter.

We give the diagram of the steel reservoir apparatus with which oxygen was liquefied in quantities sufficient to pour from one vessel into another.

A is a cylinder of oxygen gas compressed to 100 atmospheres. It is connected by a tube to the steel reservoir, *B*. From the lower end of the steel reservoir a tube with stopcock, *b*, descends. A gauge, *a*, indicates the pressure of the oxygen. It is obvious that any considerable diminution of pressure would indicate liquefaction.

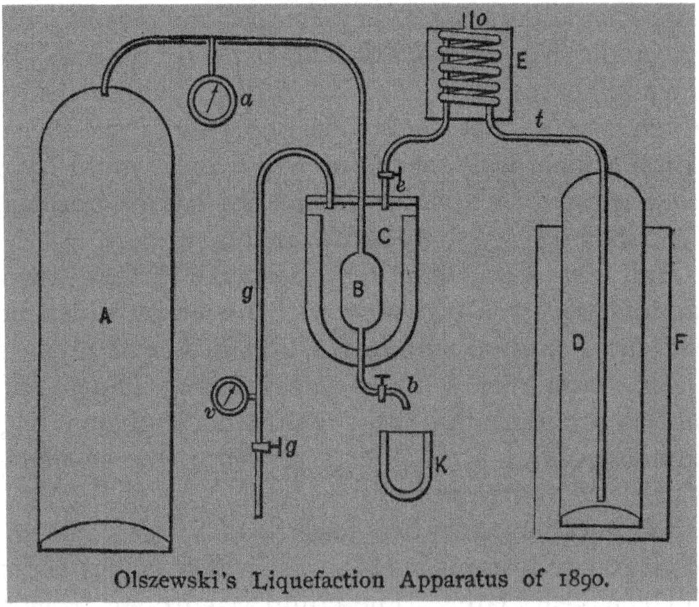

Olszewski's Liquefaction Apparatus of 1890.

The reservoir, *B*, is contained in a double-walled vessel, *C*, hermetically closed at the top. From it one tube, *g*, runs to an exhausting pump. This tube has a cock, *g*, and vacuum gauge, *v*. Another tube, *t*, runs to an ethylene cylinder, *D*. This tube has a stopcock, *e*, and is bent into a coil between *C* and *D*.

The coil is contained in a vessel, *E*, which is charged with a mixture of ether and solid carbon dioxide. A tube, *o*, leads from this vessel, which is absolutely tight, to an exhausting pump. *D* contains liquid ethylene, which is kept cold by ice and salt mixture in the outer vessel, *F*.

The oxygen under high pressure filled the steel vessel, *B*, which was quite small, of but a few ounces capacity. Here it was subjected to the refrigeration due to the liqui ethylene, cooled by exhausted carbon dioxide and ether, and also subjected to exhaustion, so as to have its temperature greatly reduced by boiling. The intense cold, which was below the critical temperature of oxygen, rapidly liquefied it under pressure, and soon the vessel, *B*, filled with the liquid. It could then be drawn off by opening the cock, *b*.

By opening and shutting the cocks the apparatus could be manipulated very readily, and the pressure gauge, *a*, and vacuum gauge, *v*, gave certain indications of the progress of operations. If the apparatus is analyzed and reduced to its elements, it will be seen to be a simplification of Pictet's apparatus of 1877, simplified by the suppression of pump circuits and by the use of compressed gases. I will be seen to be much the same as Dewar's apparatus of 1883 (page 236), and the latter expresses himself as of the opinion that the substitution of the steel reservoir for the glass tube which he employed was not a very important change.

To keep this delivery under some control, the outlet tube from the steel oxygen vessel had lateral openings. This prevented the stream of liquid from rushing out against the bottom of the vessel and driving out the contents as fast as received.

It is impossible within the limits of this work to give the entire work of any investigator. Olszewski determined many constants, by many methods, and the general abstract of his work, with table of constants determined and bibliography or list of his papers, may be found in the *Philosophical Magazine* for 1895.

For determining low temperatures he used as a matter of preference the hydrogen thermometer, and used it to standardize a platinum resistance thermometer when the temperature fell too low for the hydrogen instrum nt. But he distrusts all except the hydrogen thermometer, except under limited and defined conditions. Extrapolation he naturally suspects, and, on account of variations in specific heat as lower temperatures are reached, he has little confidence in calorimeter methods.

During his investigations he was troubled with bursting tubes. His work, like that of other investigators, was not of the safest order.

James Clerk Maxwell, one of the most illustrious physicists and mathematicians of England, had doubted the possibility of liquefying hydrogen. Faraday had not felt so. He believed that it might yet be accomplished, and expresses himself in rather uncertain phrase concerning it. Olszewski had no hopes of liquefying it in volume or as "static hydrogen." The lesson of Cailletet's production of cold by release from pressure seems to have been lost to the world, only to be successfully applied within the last five years by Tripler, Linde, Hampson and

Dewar. But without attempting to liquefy it in large volume, Olszewski tried to determine the constants of liquid hydrogen. Now, his temperatures ran so low that he was forced to use a platinum resistance thermometer, which he compared with a hydrogen thermometer, with the following result:

Temperature by hydrogen thermometer.	Electrical resistance of platinum resistance thermometer.
0° C. (32° F.)	1000 ohms.
—78·2° C. (—108·8° F.)	800 "
—182·5° C. (—296·5° F.)	523 "
—208·5° C. (—343·3° F.)	453 "

This shows the decrease in electrical resistance due to reduction of temperature which is utilized as a thermometric factor. But more is shown. The fall in electrical resistance per degree fall in temperature grows greater as the temperature descends. Thus:

	Ohms.
Between 0° and —78·2° C. the fall per degree is	2·557
" —78·2° " —182·5° C. " " " "	2·655
" —182·5° " —208·5° C. " " " "	2·692

The last figure was adopted for the extrapolation, or carrying out the scale beyond the limits of the experiment.

He found for hydrogen a critical temperature of —234·5° C. (—390·1° F.) and a boiling point at atmospheric pressure of —243·5° C. (—406·3° F.) The lowest static temperature Olszewski claims to have attained is —225° C. (—373° F.) The hydrogen temperatures were of exceedingly brief duration.

The method adopted for reaching this figure depended on the observation that if a gas is exposed to high pressure and is then cooled to a temperature not far from the critical temperature, a slow reduction of pressure will bring about liquefaction of the gas. The appearance of a mist indicated the liquefaction. The result of numerous experiments with hydrogen showed that this mist appeared always at exactly the same pressure if the experimenter started with a high enough pressure.

Thus he varied the initial pressure all the way from 80 to 140 atmospheres by 10 atmospheres at a time, cooled the compressed gas to —211° C. (—347·8° F.) and suffered the gas to expand, watching the change in pressure as it did so, and watching for the mist. This mist always showed itself at 20 atmospheres of pressure, whether the initial pressure was high or low, provided it did not range below 80 atmospheres.

If the initial pressure did fall below this point then the pressure at which liquefaction took place also fell, and, starting from initial pressure of 50, 60 and 70 atmospheres, the mist appeared at pressures of 14, 16 and 18 atmospheres respectively. All constancy was lost.

Therefore, Olszewski accepted 20 atmospheres as the critical pressure of hydrogen, and thence deduced the conclusion that hydrogen liquefying at 20 atmospheres had the critical temperature. As he could always produce the slight evidences of liquefaction at this pressure in the small glass tube, he believed that he could always produce liquid hydrogen at the critical temperature by establishing the conditions described.

The only trouble was that such a minute quantity of hydrogen was liquefied in his glass tube that it was impossible to determine its temperature. He, therefore, resorted to his steel vessel apparatus (page 224), established the proper conditions of initial pressure and temperature, slowly reduced the pressure to 20 atmospheres, and took the temperature of the hydrogen in the steel vessel.

He saw no liquèfaction, for the steel vessel hid its contents. He established the conditions which had always produced the mist in the transparent glass tube, and he relied upon the large size of the steel vessel to give enough liquid hydrogen to affect the electric resistance thermometer which he employed.

Dewar, after producing liquid hydrogen in quantity so that it could be poured from vessel to vessel, and so that its temperature could be accurately determined, comments unfavorably on Olszewski doing his work in an opaque vessel. Although, too, Olszewski's assumptions seem rather forced, and led him to too high a critical pressure figure, his results are surprisingly good, and compare well with Wroblewski's calculated ones and Dewar's presumably more accurate ones.

CHAPTER XI.

JAMES DEWAR.

Dewar's life and education—His associates—Controversies with Cailletet as to priority—Early liquefaction apparatus—Solid nitrous oxide as a refrigerant—Royal Institution apparatus—Cooling cycles employed—Laboratory apparatus—Vacuum vessels—Air as a heat conveyer—Experiments with incandescent lamps—Reflection of ether waves from vacuum vessel—Keeping power of vacuum vessels—The Dewar vacuum—Its extraordinary perfection—Analogy with population of earth—Experiment in slow diffusion of mercury vapor—Incidental production of vacuum vessels—Elasticity and strength of metals at low temperatures—Apparatus used—Elongation of metals when stressed at low temperatures—Determination of specific and latent heats of liquefied gases—Gas jet experiments—Low temperatures thus obtained—Freezing air—Large jet apparatus—Analysis by liquefaction—Liquefaction of fluorine—Liquefaction of hydrogen and helium—Experiments to show the intense cold.

James Dewar was born in 1842, in Kincardine-on-Forth. He was educated at the Dollar Academy, and subsequently at the University of Edinburgh. He acted as assistant in chemistry to Sir Lyon Playfair in the University of Edinburgh, where the former was Professor of Chemistry. He also studied in Ghent under Auguste Kekulie. He has had many honors accorded him. For sixteen years he has been Jacksonian Professor in the University of

Cambridge. He is Fullerian Professor of Chemistry in the Royal Institution of England, thus being Faraday's successor.

The list of papers by Prof. Dewar and his colleagues relating to investigations at low temperatures is a long one, extending from 1874 down to the present time, and including nearly eighty titles. His colleagues in this work comprise Professors G. D. Liveing, J. A. Fleming and Moissan. Most of the papers are by Dewar alone.

Dewar had been interested in calorimetry for a long time, and had used a vacuum vessel as an insulator in calorimetrical experiments in 1874, at the University of Edinburgh. This date was brought out in a claim of Cailletet, who thought that he antedated Dewar in this device. Had it not been for the old Edinburgh experiments, the French scientist would probably have carried his point.

An early reference of Dewar's involved him in a second controversy with Cailletet. At the 1883 meeting of the British Association for the Advancement of Science he had pointed out the advantages of a liquid of low critical pressure, such as liquefied marsh gas, for the production of intense cold. The critical temperature of this gas he put at less than —100° C. (—148° F.), with a corresponding pressure of only 39 atmospheres. He then stated that he hoped soon to approach the absolute zero by the use of this refrigerant.

Dewar set considerable store by this utterance, as he had hoped to prove by it his priority in the use of liquid marsh gas for the production of cold, which priority was claimed by Cailletet.

In 1885 he and Cailletet had a discussion or interchange of communications on the subject of the priority in the use of liquefied marsh gas, Dewar referring to his British Association remarks as published in *Nature* in 1883, and Cailletet referring to a sealed communication deposited by him with the French Academy of Sciences, dated 1881.

As a portion of his duties at the Royal Institution, Dewar had to lecture on chemistry and physics, and naturally felt called upon to show liquid oxygen to his audiences. The work of Cailletet, Pictet, Wroblewski and Olszewski was still fresh and in progress. Accordingly, Dewar had arranged a liquefaction apparatus on the lines followed by the last named investigators for exhibiting liquid oxygen to his audiences. These lines, it will be remembered, involved originally a combination of Cailletet's and Pictet's apparatus. As their work progressed, Cailletet's apparatus became less a feature of it, but Pictet's system of successful cooling cycles was preserved.

This feature is prominent in Dewar's early apparatus, and has always been retained up to the present time. Pictet set the example, which was followed in Cracow, Leyden and London, only now to be abandoned by Tripler, Linde and Hampson, who have dispensed entirely with outside refrigerants and have made air and gases supply the cold for their own liquefaction.

Dewar's early apparatus of 1883 was designed simply to liquefy oxygen in a glass tube for lecture purposes. The apparatus was arranged for projection of the gas tube by the magic lantern. It is of interest

Courtesy of *McClure's Magazine*.

Prof. Dewar in the Laboratory of the Royal Institution.

as being the predecessor of the expensive apparatus since that period installed in the laboratories of the Royal Institution. It will be seen that it differed very little from Olszewski's apparatus of 1890, except that the receiver for the liquefied oxygen was a glass tube and that no means were provided for withdrawing the liquefied gas. In any case, far too little was produced at a time to make it possible to pour it from vessel to vessel except on the most limited scale, if at all.

Prof. Dewar has been far from communicative on the subject of the liquefaction apparatus and methods employed at the Royal Institution. They are based on the Pictet system of successive cycles of cooling agents, one agent cooling the next, so as to secure several steps down the thermometric scale, the last being utilized for the gas to be liquefied. It is only very recently that a step forward has been made and the self-intensive method adopted, and in the case of his hydrogen liquefactions superadded to the Pictet cycles.

Now that the work has been done and air has been liquefied in large quantities by the expensive methods adopted and devised for the Royal Institution work, it is with a feeling of sadness that we realize that the great quantities of liquefied ethylene which excited so much admiration were not needed, and that, by the simple methods of Tripler, barrels of liquid air could have been made at relatively nominal expense.

Referring to the cut, *C* is an iron oxygen reservoir within which is the oxygen gas compressed to 150 atmospheres. *A* is the regulating stopcock by which it is allowed to flow out of the reservoir as desired. The glass tube in which the gas is liquefied is indicated by *F*, and the gas from *C* reaches it through a fine copper tube, *I*. *D* is a manometer to show the

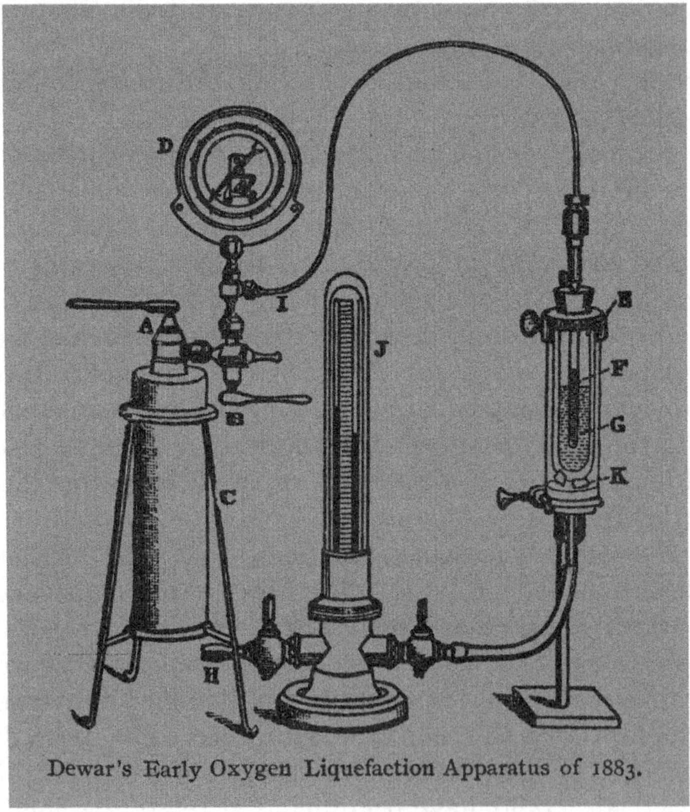

Dewar's Early Oxygen Liquefaction Apparatus of 1883.

pressure of the gas, and *J* is an air pump gauge to indicate the vacuum under which the refrigerant boils. *H* is the point of attachment of an air pump for producing this vacuum.

The gas liquefaction tube, *F*, is surrounded by another tube, *G*, also of glass, in which is liquid ethylene, liquid nitrous oxide or solid carbon dioxide. These boil in the approximate vacuum produced by the air pump. It will be observed that a third vessel, *K*, surrounds *G* and *F*, and that the exhaustion takes place from its bottom. Its top is hermetically sealed, and holes at *E* permit the cold gas from *G* to flow down the annular space between *G* and *K* to keep the temperature low.

When the pressure in the vessel, *G*, containing ethylene, is reduced to 25 millimeters of mercury, the temperature falls so low that oxygen liquefies when the manometer shows a pressure of 20 to 30 atmospheres. If liquid nitrous oxide or solid carbon dioxide is used in *G*, then the pressure of the oxygen must be brought up to 80 to 100 atmospheres to compensate for the lower temperature. Or the lower temperature produced by the last two refrigerants may be supplemented by sudden release of pressure. The cock, *B*, is adapted to effect this application of Cailletet's principle.

An ingenious suggestion is made by Dewar that solid nitrous oxide should be used instead of liquid nitrous oxide in order to prevent troublesome ebullition.

He tried the specific gravity by evaporating a measured volume of the liquid and determining its amount, and performed a number of experiments, naturally very much restricted in number and impressiveness by the exceedingly small quantity of liquid available and by its inclosure in a glass tube.

Lately, however, more has been said of the Dewar

processes of liquefaction, and details of a laboratory apparatus of his for liquefying air and other gases have been made public. In England so much interest has been excited by the work of Linde and of Hampson, and the construction and theory of their apparatus have been so freely disclosed, that it seems time for the processes of the Royal Institution laboratory to be made more public than they ever have been. Details, however, are still wanting.

It follows, therefore, that there is no possibility of exactly describing the liquefaction apparatus in question. If, however, Pictet's apparatus be taken as representing the type of a double cycle refrigerating apparatus, the following give the data of its operation for the Dewar liquefactions of five years ago.

The cooling agent of the first cycle was liquid nitrous oxide. This was compressed to about 90 atmospheres and was evaporated in a condenser jacket so as to give a temperature of —90° C. (—130° F.) Through the inner condenser chamber liquid ethylene passed. This was under a pressure of over 120 atmospheres, and was cooled by the evaporating nitrous oxide which surrounded it. The liquid ethylene, brought down to nearly —90° C. (—130° F.), was passed into the jacket of a second condenser in which it was evaporated. The intensely cold liquid, cooled still more by its own evaporation, brought about a temperature of —145° C. (—229° F.)

A tube passed through the condenser jacket in which the ethylene evaporated, and through the tube oxygen, compressed to 50 atmospheres, flowed. It liquefied rapidly, and was drawn off as required. In drawing it off at this pressure, nine-tenths of it was lost. It was another illustration of the difficulty of coping with the mechanical troubles of too high pressure. We have had occasion more than once to allude to this trouble, and Dewar's statement that he lost the greater part of his liquefied gas emphasizes what we have said about this feature of Pictet's, Olszewski's and Dewar's early apparatus. A jet of

Courtesy of McClure's Magazine.
Machinery for Operating Liquefaction Apparatus, Royal Institution.

liquid at 50 atmospheres is almost uncontrollable, and the action of a regulating cock is apt to involve some wasteful atomizing action upon the liquid.

It was with this apparatus that oxygen and other gases were liquefied by Dewar in quantities almost unhoped for up to his time, and with it liquid air was prepared for the lectures which did so much to excite public attention on the subject of the liquefaction of gases.

The apparatus was very large and heavy, and it involved the making of great quantities of ethylene by decomposing alcohol with concentrated sulphuric acid. This cost a great deal. Faraday's old laboratory became the scene of operations which recalled a machine shop rather than a scientific workshop.

Prof. George F. Barker, of the University of Pennsylvania, in visiting the scene of Dewar's work, found almost as much to admire in the dozen cylinders of liquid ethylene as in the air and gas liquefactions which it accomplished. Commenting on the strange uses to which Faraday's laboratory was put, Prof. Dewar told his friend that Faraday would have been the most delighted man in the whole kingdom had he been alive to see what was in course of accomplishment. The work was nothing but the following out of the path that Faraday pointed out, and in which he went as far as the knowledge of his time permitted.

There is no difficulty in assenting to Prof. Dewar's views thus expressed.

Simpler apparatus was constructed later, and we illustrate Prof. Dewar's small apparatus for effecting liquefactions without the use of pumps, reliance being placed on the use of cylinders of compressed gases.

In the general view of the apparatus two compressed gas cylinders are seen. The one to the right contains compressed and liquid carbon dioxide, the one on the left contains compressed and gaseous air or oxygen. The small cylinder above and in a central position contains the liquefaction appa-

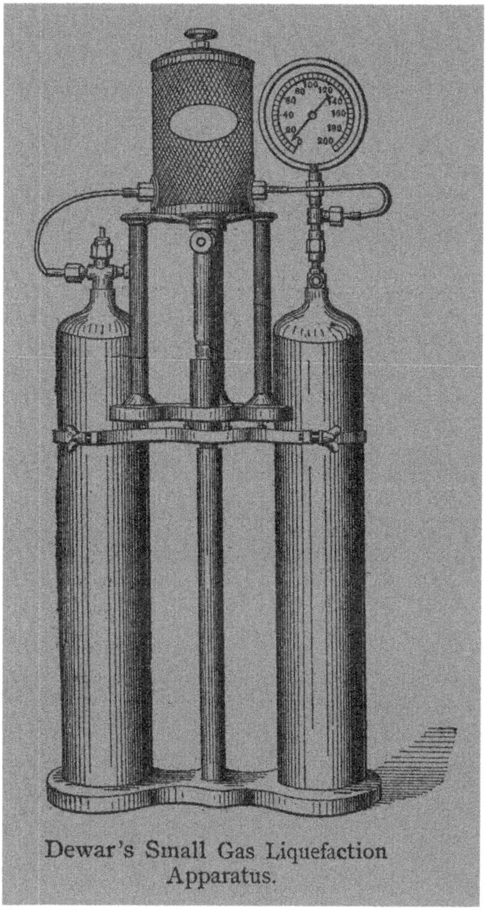

Dewar's Small Gas Liquefaction Apparatus.

ratus. It forms a very compact piece of apparatus. The next cut shows the condensing and liquefying portion of the apparatus in section.

The carbon dioxide gasifies as it escapes from the cylinder and enters the apparatus, passing in by the inlet, *B*. It follows a coil of pipe which winds around the interior of the cylinder in parallel with a second similar pipe. This second pipe communicates by the inlet, *A*, with the cylinder of compressed air or oxygen. In the sectional cut the carbon dioxide pipe is represented by the black circles, the air or oxygen pipe by the open ones. The carbon dioxide after passing through this coil escapes into the inner chamber of the apparatus and is regulated by a valve operated by the hand wheel above *C*.

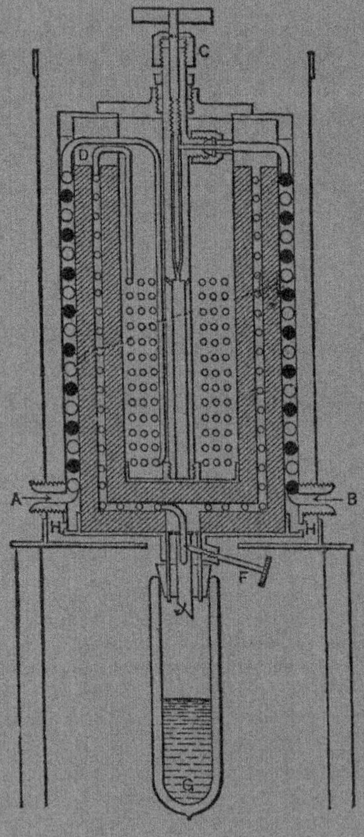

Section of Dewar's Small Gas Liquefaction Apparatus.

The air or oxygen, after going through the outer coil, and getting a preliminary cooling from the carbon dioxide coil, enters the coil in the inner chamber indicated by the triple set of small open circles. Here it circulates around through a great length of pipe and is further cooled by the expanding carbon dioxide, then goes through a third coil, intermediate between the outer coil and the inner chamber, and escapes, regulated by the valve, *F*. It liquefies and collects in *G*.

In operation the carbon dioxide solidifies so that the gas is cooled by the solidified carbon dioxide gas.

This apparatus was operated without exhaustion, the natural evaporation of the carbon dioxide giving a reduction of temperature to —79° C. (—110·2° F.) The tubing is of copper, to secure good heat conduction and consequent rapid cooling. The rest of the refrigeration is due to the expansion of the oxygen. It is well to start with this gas compressed to 150 atmospheres and to utilize it down to a pressure of 100 atmospheres. The liquid air or oxygen begins to drop in about fifteen minutes. The intensely cold expanded and unliquefied gas rises among the coils and cools them still more, so as to obtain a regenerative action. The apparatus will make 100 cubic centimeters (about six cubic inches) of liquefied oxygen in an operation.

The spheroidal state has been somewhat fully treated in an earlier portion of this work. The original investigators of the phenomena of the liquefaction of gases never imagined how important a part it would play in facilitating their manipulation. Thanks to it, the hand can be immersed in liquid air. Liquid air rests quietly in a tin dipper, and the length of time for which it remains in the open air in a common vessel is in many cases due to its taking the spheroidal state.

But liquefied gases do evaporate rather rapidly in

the air, and a great desideratum was some kind of a vessel that would hold them without the rapid loss experienced under ordinary conditions. Liquefied gases volatilize and disappear because they receive heat from surrounding objects and from the atmosphere. Early in his scientific work Dewar recognized that it might be possible to make this loss very much less, utilizing a vacuum as a non-conductor.

The properties of a vacuum in intercepting the transmission of heat are utilized in what are known as Dewar's bulbs for holding liquefied gases. Air is often spoken of as a good insulator, and such it is. Absolutely quiet air is nearly as good an insulator as a vacuum.

But the trouble is that air cannot be kept still, and if it is free to move, its mass, under the influence of heat, travels back and forth and carries heat with it, and thus by convection destroys the heat insulation of objects it is in contact with. Among objects in everyday use the incandescent lamp may be referred to as one in which a vacuum is utilized. A very considerable proportion of the efficiency of an incandescent electric lamp is due to the vacuum within the bulb. The vacuum is not only useful in preserving the carbon from combustion—a filling of the bulb with nitrogen gas would do this—but it keeps cold gas of any kind from coming in contact with the film and thereby cooling it.

The incandescent lamp illustrates so admirably the heat insulating properties of a high vacuum that some experiments may here be cited which show the effect of filling the bulb of an incandescent lamp with various gases as contrasted with having it empty.

As the vacuum protects the film of an incandescent lamp from cooling, so does it protect a mass of liquefied gas from heating. Dewar's very elegant invention is illustrated by an appeal to the other end of the thermometric scale from that occupied by liquid air.

In the *Philosophical Magazine* of 1894 we read that Blenkroode filled three incandescent lamps with carbon dioxide, coal gas and hydrogen respectively. A fourth lamp of the regular construction with a high vacuum existing in the bulb was added to the series, they were placed on a lighting circuit, and a piece of phosphorus was placed on top of each one. On passing a current through them, the vacuous lamp was the brightest, the presence of the gases chilled the other carbons, and the phosphorus was ignited in the following order: first, on the carbon dioxide lamp; second, on the coal gas lamp; third, on the hydrogen lamp, the regular lamp being the last on which the phosphorus ignited. The lamps varied in brightness in the same general order, the regular vacuous bulb lamp being by far the brightest. This illustrates the utility of a vacuum as a heat insulator.

In the case of the incandescent lamp the problem is to maintain the heat of an incandescent body in the vicinity of relatively cold objects. In the case of liquid air and gases the reverse has to be effected. A very cold body is to be prevented from receiving heat from surrounding matter. But, as is so often the case, opposites here come together, and the same means which will keep the film in the lamp from losing its heat will prevent liquid air from losing its cold, if such an expression may be allowed.

A double-walled glass vessel in a measure preserves liquid gases from evaporation. The inclosed air acts as an insulator, but, by convection, carries heat from outer vessel to inner one. A triple-walled glass vessel is still better, as it gives two spaces filled with air. The earlier experimenters used double-walled vessels for another purpose. They found that liquid gases in a single glass vessel caused ice to rapidly form upon its outer surface, so that the contents were hidden, as the ice was white and opaque. They employed a double vessel and placed some drying agent between the two vessels, on the bottom of the outer one, to keep the air between them dry, so that no such ice could form. We have seen how in his early work Dewar used this device and others did the same.

The Dewar vacuum bulb consists of a double or treble walled glass vessel, with the space or spaces between the vessels hermetically sealed and with a nearly perfect vacuum therein. The conditions in such a vessel are that the liquid in the interior one receives practically no heat. Glass is so poor a conductor that it conveys only slight traces by conduction. The liquid receives none by contact with the air above it, as it is overlaid by the intensely cold gas evolved from itself. The vacuum surrounding it cuts off any heat from warm air coming against the sides of the containing vessel. Almost the only heat it can receive is that imparted by ether waves or, popularly speaking, by radiant heat.

Ether waves of this description are such as we feel when we hold the hand near the bulb of an incandescent lamp when hot and giving light. They pass through glass with little loss. If the glass of the inner bulb, the one containing the liquid air or liquefied gas, were coated with some bright opaque substance that would reflect these waves, a further economy would be obviously effected.

This was done for the glass bulbs by coating the surface of the inner bulb with silver. The bright metal reflected the ether waves, and a better effect in preserving the gas was the result.

Then a still simpler treatment was discovered. A little mercury—a very little suffices—was left in the vacuous space outside the containing bulb. When liquid gas was put into the bulb, it chilled it and condensed a mirror of mercury upon its outer surface, which reflected the heat waves. When the liquid gas was removed, the mercury disappeared again.

Direct tests showed that a vacuum preserved the air about five times longer than would air. The following figures are given:

Relative Volumes of Liquefied Gases Evaporating from Double Bulbs.

Liquid oxygen,	vacuum space,	170	volumes.
" "	air "	840	"
" ethylene,	vacuum "	56	"
" "	air "	230	"

If the silvering process is applied, the influx of heat is reduced to about one-thirtieth of what it is with an air space, or, in round numbers, $3\frac{1}{3}$ per cent.

Three dry air spaces, one outside the other, only reduced the influx to 35 per cent. of what it was with a single space.

It is interesting to find that Prof. Dewar had

used metallic vacuum vessels in 1873 in calorimetric experiments, which he describes in a paper read before the Royal Society of Edinburgh, and printed in their *Transactions*, vol. xxvii.

Courtesy of McClure's Magazine.

Various Shapes and Modifications of Dewar Bulbs, and Liquefied Gas Containers.

Various shapes can be given to the bulbs, and several are shown in the cut. The mercury silvering process is not always employed, as it may be desirable to have the liquid visible, and the deposition of mercury on the glass cuts the liquid off from view.

For some reason, vacuum vessels deteriorate. The

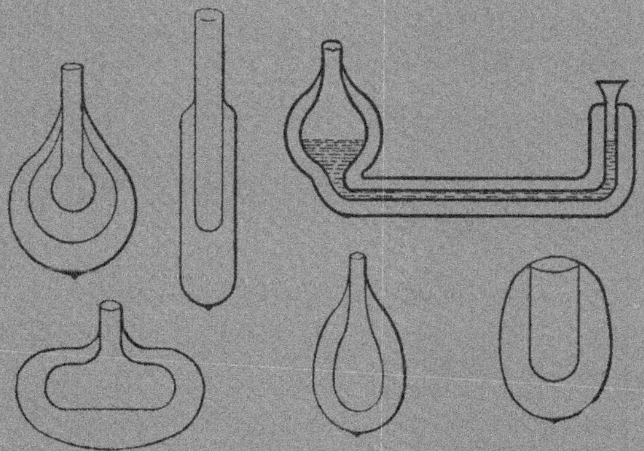

Sectional Views of Different Forms of Dewar's Bulbs.

vacuum cannot well be supposed to diminish, and no satisfactory reason can be given for the change.

Prof. Dewar adopted, for the exhaustion of his vacuum vessels, the principle of the Torricellian vacuum combined with that of freezing mercury vapor.

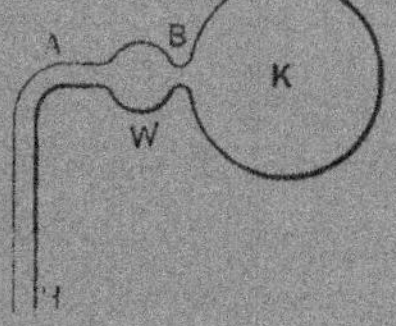

Production of Dewar Vacuum.

Suppose that the drawing represents a glass bulb, *K*, for production of a vessel in which a Dewar vacuum, as it may be called, is to be produced. The large bulb is the one which will eventually form the vessel. From the small bulb, *W*, the tube, *H*, descends a distance of over 30 inches.

To simplify the description, the single outer bulb alone is shown. The inner bulb is not represented in the drawing. It must be supplied by the reader's imagination.

The whole affair, tube and bulbs, is filled with mercury while inverted, exactly as in filling a barometer tube. By heating, or by some other manipulation, any air present may be expelled, for mercury, mobile as it seems, invariably holds bubbles of air imprisoned when it is poured into a long tube. In filling barometers, several methods of getting rid of the air are employed, boiling the mercury being one of the best. Barometers thus treated are said to have "boiled tubes."

The long tube with the large and small bulb, being filled with mercury, is reversed in position, with its lower end immersed in a cistern of mercury. The mercury descends until it stands at a height of about 30 inches. By a little inclining of the tube, any mercury remaining in the bulbs can be made to enter the tube, or a little may be left there as a silvering agent. In the bulbs the Torricellian vacuum now exists. It would be a perfect and absolute vacuum except for the presence of mercury vapor. A blowpipe flame is applied at the outlet, *A*, of the small bulb, the tube melts together, and the two bulbs are removed hermetically sealed. A trace of mercury vapor is still in them.

The next operation is to chill the small bulb by wiping it with a piece of cotton dipped in liquid air. As this touches the glass, the mercury vapor is frozen solid and is deposited on it and forms a mirror. This mercury is derived from the vapor which exists in the bulbs. A sufficient freezing removes almost every trace of vapor, and the mercury vapor is removed from the large and from the small bulb.

Keeping the bulb cold with liquid air, the small neck between the bulbs is sealed off by the blowpipe flame, and the large bulb has now within it the most complete vacuum known. It is all but absolute. Some infinitesimal traces of mercury vapor are present. It responds to the most severe electrical tests for vacua.

While a sufficiently long exposure of the small bulb to the absolute zero, were such attainable, might make the vacuum absolute, the difference between it and the Dewar vacuum would be infinitesimal.

The calculated pressure of mercury vapor at the temperature of melting ice is expressed by the decimal 0·000,126 millimeter of mercury. The reference is to a barometric column of mercury which has a normal length of about 760 millimeters. Therefore, the above decimal expresses one six-millionth of an atmosphere, certainly low enough for almost any purpose. But on lowering the temperature to —180° C. (—292° F.) by sponging the outer bulb with liquid air, the pressure of the mercury vapor falls to the figure 0·000,000,003 millimeter, or two and a half millionths of a millionth of an atmosphere. In powers of ten it would be expressed by 25×10^{-13} of an atmosphere.

If a bulb of identical size were filled with mercury vapor at atmospheric pressure, it would, therefore, contain two and a half million million times as great a weight of mercury. If it were filled with air at atmospheric pressure, it would contain in round numbers one-fiftieth the above weight of air, or eighty thousand million times the weight of the mercury in the Dewar vacuum.

Amazingly small as this quantity is, we can obtain some concrete idea of it from the population of the world. This may be taken at about one thousand millions. If then we had one thousand earths, and removed from them all of the human inhabitants except three, they would represent three-millionths of a millionth of the original population of our thousand worlds.

Prof. Dewar seems amply justified in maintaining that the vacuum he produces is higher than any of which man had ever yet dreamed.

The rate at which mercury is thus deposited has been investigated. All that was necessary was to apply the cooling process to a vacuum bulb containing a globule of mercury. The latter supplied more mercury vapor as fast as, or nearly as fast as, it was deposited on the glass. The time of cooling was taken, and then the bulb was broken and the mercury weighed. The area over which the mercury was deposited being known, the data are reduced to mercury deposited on a given area in a given period.

In ten minutes two milligrammes of mercury were deposited per square centimeter of surface. This gives a rather interesting figure. These two milligrammes of mercury represent enough vapor to saturate in the Torricellian vacuum no less than twenty liters or about twenty quarts capacity. A globe big enough to hold this quantity, if exhausted by the Torricelli process, would contain just about two milligrammes of mercury vapor, and ten minutes' cooling by liquid air sponging would remove this from the globe.

Remembering that two milligrammes are equal to about three hundredths of one grain, and that twenty liters are equal to about twenty quarts, and that a twenty-liter globe would hold seven pails of water, we again have a concrete example of the effect of removing mercury vapor from a Torricellian vacuum. It also gives us an idea of how near perfection a Torricellian vacuum is, and of what is gained by the freezing process applied to it.

In scientific work one must always be on the watch for side issues. New and interesting facts constantly come out by accident, or are suggested in investigations having widely different ends in view. An interesting example occurs in the freezing of the mercury vapor in the bulbs we describe.

The cut shows an apparatus designed to show the slowness with which mercury gas diffuses through a long, slender glass tube. Two bulbs, a large and a small one, are connected by a capillary tube. The latter in the experiment as executed by Dewar was 2 millimeters (0·08 inch) in diameter and 50 millimeters (2 inches) long. A little globule of mercury is in the smaller tube. A Torricellian vacuum is produced by the process already described, and the tubes are sealed up so as to maintain it within their interiors.

The cotton wad, *A*, wet with liquid air, is applied to the large bulb, and a mirror at once forms where the same is applied. All the mercury gas in the large globe deposits there, and, on touching another portion of the glass, no mirror shows itself. The mercury gas diffuses with such extreme slowness through the capillary tube that the latter for a while acts almost like a valve to shut off the communication between the two bulbs.

If now the bulbs are inclined so that a little mercury runs into the large one, then, on applying the sponge elsewhere, a new mirror is at once formed.

Such are the Dewar bulbs, one of the most valua-

Dewar's Experiment of Freezing Mercury Vapor in Connected Bulbs.

ble of the mirror devices in connection with our subject. These bulbs and the spheroidal state are what enable liquid air and gases to be handled almost as if they were so much water. Certainly, the ease of handling is comparable to the case of a volatile inflammable liquid, such as benzene or ether.

It is interesting to observe that sometimes the principle has been applied in a sense unconsciously. Thus, for the production of low temperatures, a vessel is often surrounded by another one containing a liquefied gas. The joint between the vessels is hermetically tight, and the liquefied gas in the space between is reduced in temperature by the application of exhaustion, thus making it boil.

Although this vacuum is applied simply to reduce temperature, one of its actions is to make the combination constitute approximately a Dewar vacuum vessel.

We now pass to some determinations of data at low temperatures, giving as required illustrations of the apparatus employed by Dewar and his associates. Much ingenuity was required in carrying out some of these determinations, but they were made possible by the ample facilities for the production of liquid air and liquefied gases. Had the experimenters, relatively speaking, had such quantities of liquid air at their disposal as have been produced in New York city by Tripler, their tasks would have been still easier.

The strength of metals and their rigidity are greatly modified by extreme cold. It is easy to show this in a crude way. Thus a spiral of soft metal, such as solder, an alloy of lead and tin, may be drawn out into a straight line by suspending a very small weight by it. But if cooled to the temperature of boiling oxygen or thereabout, it will support a weight fifteen or twenty times greater than before, without being drawn out of a spiral, and will spring like a watch spring.

This experiment gives us an explanation of Tresca's flow of solids. He found that, under great pressure long maintained, metals would flow like a very thick liquid, but very slowly. The soft metal, which is so easily straightened out, may be supposed at our everyday temperatures to exist in a state of semi or partial, perhaps much less than semi-fusion. The same becomes fully solid when cooled by liquid air.

The same spiral will vibrate like a steel spring when intensely cold. At ordinary temperatures it is almost devoid of elasticity.

A tuning fork or bell made of such metal will not ring at ordinary temperatures, but when chilled the elasticity is increased so that the metal becomes sonorous, the bell rings and the tuning fork sounds as if of steel or of bell metal.

As an analogy at more familiar temperatures, we may refer to iron or glass. Either of these is rigid and elastic, but when heated becomes soft gradually, not melting at once, but passing through a slow change extending over many degrees range of temperature, and gradually approaching fluidity. We may assume that such metals as lead or tin at ordinary temperatures are undergoing a change of state, and are approaching fluidity. The only trouble with this view of the case is that, when such metals do melt, the melting is sudden, and is done within a very small range, perhaps less than a degree.

If one of two tuning forks which are in perfect unison is chilled in liquid air, and the two are sounded, they are found to be no longer in unison. The colder one is of higher pitch than before, because the intense cold has made it more elastic than it was.

The difficulties of determining the strength of materials when cooled to the liquid air temperatures have been quite successfully overcome. The cut shows the general plan of apparatus used by Prof. Dewar for determining the tensile strength of materials. As the piece should be of sufficient size to insure absence of flaws of any kind, the apparatus must be powerful. Metals increasing in tensile strength as cooled, the jaws of the apparatus which hold them have also to be cooled. Otherwise, the portions of the test piece near the jaws, being warmer than the rest, would be weaker than the rest, and the sample would break there, and invalidate the test.

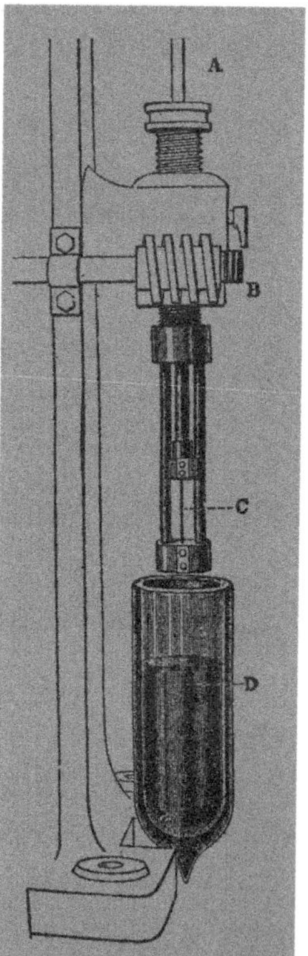

Apparatus for Determining Tensile Strength at Low Temperatures.

In the cut, *D* is a silvered vacuum vessel of liquid oxygen, *C* is the wire to be tested, *A* is a steel rod which runs to a set of multiplying levers which produce the breaking strain. At *B* is an arrangement for determining the amount of extension of the wire before breaking. When the test is to be made, the lower part of the apparatus is immersed in the liquefied gas, and the strain is applied.

If the heavy apparatus strikes the vessel, the glass will break, and an expensive piece of apparatus will be destroyed, and the liquefied gas will be lost. For this reason the apparatus has to be solidly constructed, so as to be secure from shaking or jarring under the heavy strains and from the sudden breaking of the sample under test.

As a rule, Dewar used wires about one-tenth of an inch in diameter and two inches long. He gives the following table of his results. We quote it as published in the *Transactions* of the Royal Institution. The work was published in 1894.

Breaking Stress of Metallic Wires in Pounds, 0·098 *Inch Diameter, at* 15° *C.* (59° *F.*) *and* —182° *C.* (—295·6 *F.*)

	15° C. (59° F.)	—182° C. (—295·6° F.)
Steel (soft)............	420	700
Iron..................	320	670
Copper...............	200	300
Brass.................	310	440
German silver.........	470	600
Gold..................	255	340
Silver................	330	420

The great increase of strength is due entirely to the reduction of temperature. When the wires are restored to their original temperature, the increase in strength disappears.

The inhabitant of a world where the temperature approximated the absolute zero would have much stronger iron and steel with which to build his bridges, and he might make his watch springs out of pewter and his bells out of tin.

With the same apparatus the breaking strain under longitudinal tension of test pieces of various cast metals was tried. The samples were all cast into shape. They were two inches long, they had hemispherical ends one-half inch in diameter and a central cylindrical section two-tenths inch in diameter. This gave a shape somewhat like a dumbbell.

The ends were received by cavities in the special steel end blocks in the testing machine, in which blocks hemispherical cavities were turned out to fit them. Although much discordance obtained among the results, the same general principle held as for tensile strength of wire. The chilled metals were stronger than at ordinary temperatures. The table of results we give here:

Breaking Stress of Cast Metallic Test Pieces in Pounds, 0·2 Inch Diameter, at 15° C. (59° F.) and —182° C. (—295·6° F.)

	15° C. (59° F.)	—182° C. (—295·6° F.)
Tin	200	390
Lead	77	170
Zinc	35	26
Mercury	0	31
Bismuth	60	30
Antimony	61	30
Solder	300	645
Fusible metal (Wood's)	140	450

The abnormal results with zinc, bismuth and antimony are striking. These three metals are highly crystalline, and in this feature perhaps some explanation may lie hidden.

The elongation results were not considered of any high degree of accuracy, but certain points were brought out by them. Thus tin and lead, at ordinary temperatures, elongate to the same extent before breaking; but after reduction of temperature, tin hardly stretches at all, while lead is as ductile as ever. Solder and fusible metal stretch less at the lower temperature. Steel has its elasticity only slightly changed by refrigeration. Lead, tin, iron and ivory balls, when refrigerated, are increased in elasticity and bound higher than be-

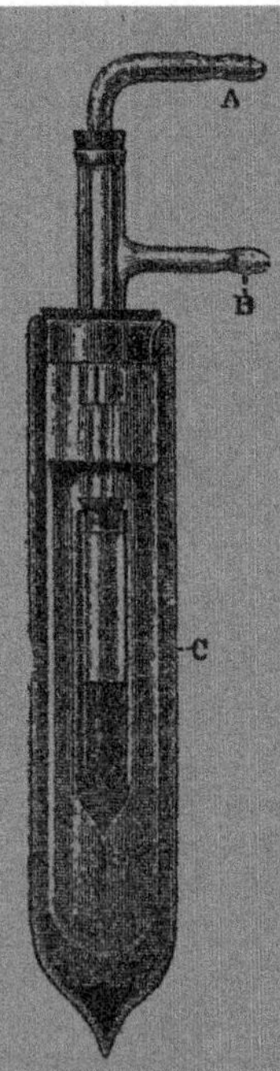

Apparatus for Preserving and Freezing Liquid Air.

ore when dropped upon an iron an- il. The cooled lead ball has a much maller distortion produced where it trikes the anvil that it would were it incooled. The area of the distortion urface is about one-ninth what it vould be in a sphere of the same netal and size at ordinary tempera- ures.

The cut shows how air, when lique- ied, can be preserved practically vithout evaporation, although at the xpense of the evaporation of other iquefied air. Two vacuum tubes are ised, placed one within the other, as hown in the cut. The inner one onnects with a tube, *A*, the outer ne, *C*, with a tube, *B*. The sample f liquid air to be preserved intact is laced in the inner vacuum tube. The uter tube contains enough liquid air o completely immerse the inner tube. By india rubber perforated stoppers, the necks of the vessels and of the tube, *B*, are closed airtight, except for the passage through them of the tubes, *A* and *B*. All heat received is cut off from the inner tube. The liquid air in the outer tube boils off slowly, and the liquid air in the inner tube is effectively preserved. If exhaustion be applied to *A* and *B*, the air in the inner tube freezes to a jelly-like mass.

The apparatus shown in the cut was the apparatus used for determining latent heat of evaporation or the specific heat of a liquefied gas. The first requirement in thermic work is to have a mass of the liquid under perfect control. It must be so placed that it will be permanent, and not evaporate. This condition is brought about by the arrangement shown in the cut, practically a duplication of what has just been described, with some additional features.

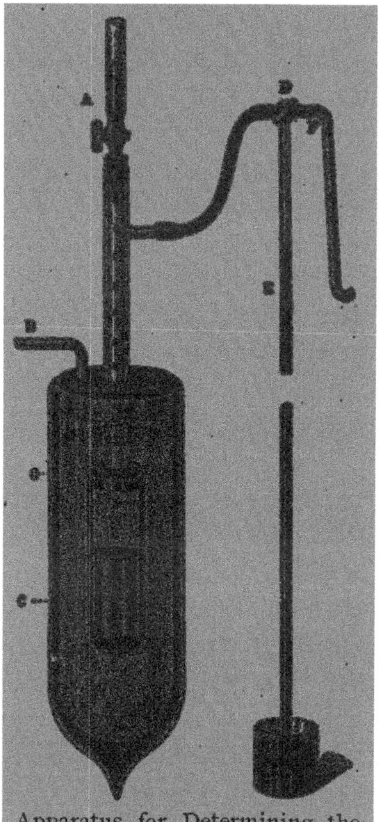

Apparatus for Determining the Latent Heat of Evaporation and Specific Heat of Liquefied Gases.

There is an outer vacuum vessel, *G*. In it is placed the refrigerant, liquid air, oxygen or such liquefied gas as may be chosen. This vessel is corked, and a second vacuum vessel, *C*, is maintained concentric with it and immersed in the refrigerant.

Latent heat of evaporation is determined by adding a known quantity of heat to the liquid and determining the quantity of gas evolved. Enough heat must be imparted to bring about evaporation, which heat may be imparted by dropping mercury into the liquid, as shown in the cut. Sometimes a piece of platinum, glass or silver is used. The weight of the substance added, its specific heat and its temperature being known, the quantity of heat imparted is calculated from these data. The gas evolved is collected, and its weight being known, the data are given for determining the latent heat of gasification or of evaporation. The gas evolved is measured, and its weight is calculated from its known specific gravity.

We now know the amount of heat added, and we know the amount of liquid which it has converted into gas. This gives the data for calculating the latent heat of evaporation. To determine the specific heat, we have to ascertain the quantity of heat required to change the heat of a given amount of the liquid from one known temperature to another. These known temperatures are the boiling points at specified pressures. When such a pressure is produced, the temperature of the boiling point at that pressure is reached. The following describes the execution of a determination of latent heat of a liquefied gas:

The capacity of the vacuum vessel, *C*, being known at given heights, it gives the quantity of liquid contained in it.

At *D* is a three-way cock. When turned in one direction, it cuts off the tube, *E*, and establishes communication between *F* and the vessel, *C*. In another position it cuts off the tube, *F*, and connects *E* with the liquid gas vessel, *C*. The tube, *F*, leads to an air pump. The tube, *F*, being put in connection with *C*, exhaustion is applied until a vacuum of about one-half an inch of mercury is produced. This fixes a temperature for the liquid gas—the boiling temperature at that pressure—which temperature is known. The stopcock, *D*, is turned so as to shut off *F* and bring *E* into communication with *C*.

The height of column is the vertical distance from the level of the mercury in the cistern to the level of that in the tube. Heat is now imparted by dropping mercury into *C* until the column of mercury in *E* sinks to the level of that in the cistern.

Now heat enough has been imparted to raise the temperature of the liquid gas from its boiling point at one-half an inch pressure to its boiling point at atmospheric pressure, the latter being taken for each experiment from a standard barometer. The quantity of liquid gas thus raised in temperature being known, the data for determining specific heat are known.

The mercury dropped into the liquefied gas in *C* needs particular management. It has a propensity for forming a stalagmite as it falls into the intensely cold liquid, and this must be prevented by dropping the mercury into different parts of the liquid. Another difficulty is the splashing of the liquid as the mercury falls into it.

The latent heat of evaporation of liquid oxygen is

about the same as the latent heat of melting of water, or 80 units, or the heat required to vaporize one part by weight of liquid oxygen would raise the heat of the same weight of water through 80° C. (144° F.)

The behavior of a jet of gas issuing from a state of high compression may be studied by such apparatus as that shown in the next cuts. The apparatus was used by Dewar. In each piece is recognizable a vacuum tube with coil.

In the first cut, *C* is a vacuum vessel which contains a coil of tubing about 0·2 inch diameter. The vessel in the experiment is filled with a refrigerant such as liquid air. The tube is of silver or of copper, so as to be a good conductor of heat. At the end, *A*, is a minute aperture.

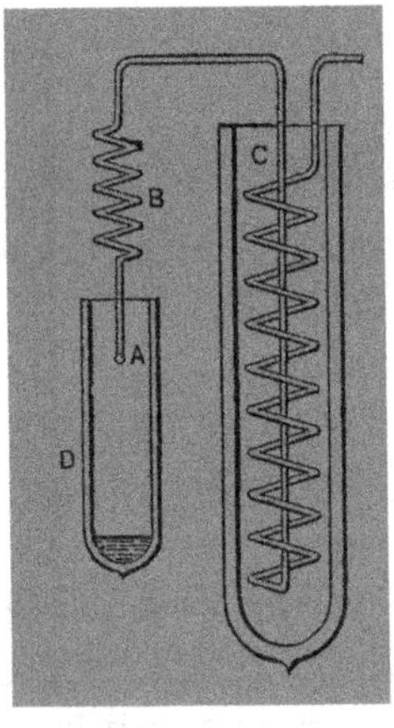

If oxygen gas at a pressure of 100 atmospheres is driven through the tube, escaping through the aperture, having previously been cooled in the tube, *C*, to a temperature of —79° C. (—110·2° F.), a liquid jet is just visible. The conditions here are not nearly so extreme as with Pictet in his experiment of 1877, in which a pressure of 270 atmospheres was used. Dewar believes that one reason Pictet required so high a pressure was on account of his stopcock being massive and being outside of the refrigerating apparatus. It is also quite possible that Pictet used a higher pressure than was really needed.

With air driven through the tube instead of oxygen, 180 atmospheres are needed for liquefaction, and with a reduction of temperature to —115° C. (—175° F.) liquid air can be collected in vacuum vessel, *D*. This reduction is effected by applying exhaustion to the carbon dioxide in *C*. Or adhering to the natural evaporation temperature of carbon dioxide (—79° C., —110·2° F.), a pressure of 200 atmospheres at that temperature liquefies air. Naturally, Dewar found that the high pressure interfered with the collection of the liquid. An interesting point he speaks of is that the collection of liquid air can be increased by directing the jet against the tube above the hole. This to some extent brings out the self-intensive principle of Tripler's, Linde's and Hampson's apparatus. By putting in a greater length of tube, as by making a coil, *B*, the efficiency is increased. This is undoubtedly because the cold gas rising produces self-intensive action. An egg-shaped vessel acts in the same way. Dewar terms it the cold regenerative process, citing Coleman, Solvay and Linde as users of this principle.

The next cuts show modifications. In the first cut the pipe is coiled around an inner vacuum tube to get better insulation from heat. The inner tube is 9 inches long and 1½ inches in diameter. Over the end of the metal tube a glass tube is slipped which stops the splashing about of and loss of the liquid air. It is evident that with such an apparatus the cold regeneration would be very well carried out. The tube is coiled in a very restricted space, and the ascending excess of unliquefied air and of evaporated air at a very low temperature comes in contact under conditions of high efficiency

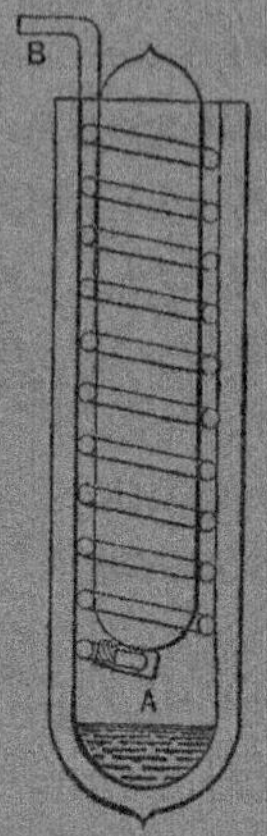

with the metal coil. It is not surprising to hear that with a pressure of 200 atmospheres liquid air begins to collect in about seven minutes. The apparatus suggests one of Tripler's early coils.

Another disposition is shown in the last of the cuts, where the gas pipe is coiled disk fashion, leaving room in the center for introduction of a glass tube, *C*, in which samples can be placed which it is desired to subject to low temperature. The glass cap to prevent splashing is seen in this cut also.

These simple jet experiments are a good introduction to a study of the self-intensive apparatus, whose use has excited so much interest, both popular and scientific.

Taking the critical temperature of hydrogen as 31° C. absolute or —242° C. (—403·6° F.), it will be seen that the temperature of boiling air (—194° C., —317·2° F.) is well above it. —194° C. is 80° C. absolute, so that boiling air may be said to be two and one-half times hotter than liquid hydrogen at the critical point. It is not clear that this is a perfectly fair way of looking at it, however.

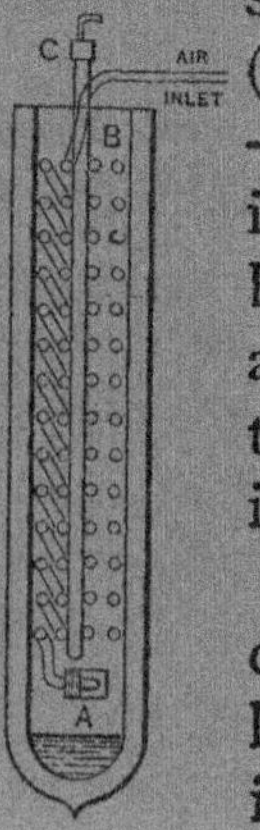

Wroblewski and Olszewski had concluded that hydrogen had an abnormally low critical pressure. Wroblewski gave it a critical pressure of only 13·3 atmospheres, which is about one-fourth that of oxygen. The only trouble, therefore, should be to get the temperature down. Dewar attempted to liquefy a mixture of hydrogen with two to five per cent. of air, and says that he obtained solid air together with a very volatile liquid of low density which he was not able to collect in a separate vessel. Olszewski longed for a gas intermediate in its critical point between air and hydrogen, to get what has aptly been termed static hydrogen, or hydrogen liquefied in quantity.

Accepting Dewar's view that hydrogen at 80° C. absolute is two and one-half times as hot as it is at its critical temperature, and taking air at two and one-half times its critical temperature, we should find that the liquefaction of hydrogen from the initial temperature of boiling air would be equivalent to the liquefaction of air from 60° C. (140° F.) or 333° C. absolute. This figure is thus reached: The critical temperature of air is taken at —140° C. (—220° F.) This reduces to 273—140 = 133° C. absolute. Two and one-half times 133 are 333, which is the absolute temperature, two and one-half times greater than —140° C. (—220° F.), and 333° C. absolute is equal to 333—273 = 60° C. (140° F.) It is possible to liquefy air by the jet method from a still higher temperature than this. Dewar found that starting with air at an initial temperature equal to that of boiling water, he could liquefy air in seven minutes by the processes described.

It would, therefore, seem as if hydrogen at the initial temperature of the boiling point of air should be liquefiable by the process which liquefied air from the initial temperature of boiling water.

Hydrogen was cooled a few degrees below this

point, to —200° C. (—328° F.) and was driven through a fine aperture under a pressure of 140 atmospheres, but without result. A very little oxygen, some few per cent., was mixed with the hydrogen, and a liquid was obtained which contained hydrogen in solution, but was principally oxygen. It gave off hydrogen and oxygen in explosive proportions.

The experiment was now tried with the regenerative coil in the first figure of the cuts, page 264. The escaping gas cooled the coil, *B*, and the regeneration brought about, apparently, a liquefaction of hydrogen. A liquid jet could be seen after the circulation had continued for a few minutes, and a liquid in rapid rotation in the bottom of the vacuum tube, *D*, could be discerned.

The difficulty of recognizing a volatile, highly mobile liquid, formed under such conditions, and so very evanescent in duration, cannot be too strongly insisted on. A stream of gas was rushing out of an orifice at fifty times the pressure of steam in an ordinary boiler, a portion of it liquefied for a very brief period, and then gasified. The violence of the operation would at least tend to confuse quiet observation.

Dewar states that, owing to the low specific gravity of the liquid and the rapid current of gas, the latter impelled by a pressure of 140 atmospheres, or about one ton pressure to the square inch, none of the liquid in question accumulated. "Static hydrogen" was almost produced, the liquefaction was destined to be soon accomplished, and in its proper place (page 280) will be found described.

The jet system of cooling by impingement has in

several places been alluded to. Cailletet in early days, unable to conceive of the possibility of using liquefied gases by the gallon as refrigerants, suggested the use of ethylene jets for cooling. It was the *chalumeau du froid*, or cold blast blowpipe, of Thilorier.

Dewar tried his hydrogen jet as a refrigerant. Liquid air and liquid oxygen were successively placed in the bottom of the vacuum tube, *D*, so as to cover the jet. In a few minutes, in each of the two cases, about 50 cubic centimeters (3 cubic inches) of the air and oxygen respectively were solidified into hard, white solids like avalanche snow.

When the air was solidified by evaporation *in vacuo*, the product was a jelly; but in the experiment just described, the cold was so much more intense that oxygen-ice and air-ice were produced. The solid oxygen had the characteristic bluish color of the liquid oxygen. Light reflected from it showed in the spectroscope the characteristic bands shown by light transmitted through liquid oxygen.

In the description of these experiments the Joule-Thomson effect (page 297) was taken no cognizance of. All was treated by Dewar as examples of cold regeneration, not of internal intensification. There is a very open question as to how important a role the Joule-Thomson effect really plays in these cases. Hydrogen, it will be remembered, does not present the effect, but the reverse. On escape from pressure under what may be termed Joule-Thomson conditions—conditions adapted to bring out the Joule-Thomson effect—its temperature rises. In the experiment, as described by Prof. Dewar, the hydrogen liquefaction is described as due to simple cold regeneration. It would seem as if it was rendered less powerful by the heating, or, as it may be termed, by the negative Joule-Thomson effect found to exist with hydrogen, unless, as Dr. Onnes

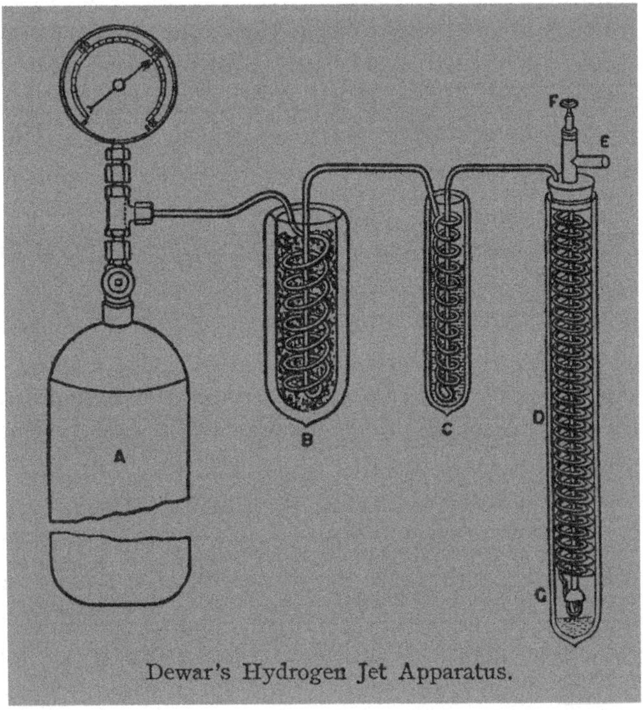

Dewar's Hydrogen Jet Apparatus.

believes, the negative effect is reversed at low temperature.

The illustration shows the general scheme of Dewar's more elaborate apparatus for cooling hydrogen by its own expansion. *A* is a cylinder charged with hydrogen under high pressure. *B* and *C* are vacuum vessels, each inclosing a coil of the gas delivery pipe. *B* contained solid or liquid carbon dioxide. The vessel was closed and its interior kept under exhaustion so as to lower the temperature. *C* contained liquid air. *D* is the self-intensive coil terminating at *G*, where there is a pinhole aperture. The first evidence of the intense cold in the freezing of air to a hard solid led to the erection of a very powerful apparatus, by means of which the liquefaction of hydrogen was effected.

This liquefaction is the last great achievement in the field we are studying. The subject, therefore, will be dropped for a few pages in order to preserve the chronological relations.

Air is always contaminated with carbon dioxide gas, and the small quantity normally present, four parts in ten thousand, which, however is subject to considerable variation, suffices to produce a turbidity in the liquefied product. Oxygen made as it usually is, from potassium chlorate by ignition, contains traces of chlorine, and this tends to produce turbidity in the oxygen when liquefied.

There are cases where in a mixture of gases one constituent liquefies while the other solidifies. It is possible to purify a gas from some mixtures by liquefying the mixture and filtering. In lecture experiments with liquid air, it is usual to filter the liquid in order to procure transparent samples to show the faint blue color.

Gases, however, sometimes dissolve in other liquefied gases, just as they do in water. Soda water is a solution of carbon dioxide gas in water. Thus liquid air dissolves hydrogen. It is found that as much as twenty volumes of gaseous hydrogen may be dissolved in one hundred volumes of liquid air. This, however, is not a large quantity. It must be remembered that the one hundred volumes of liquid air give when gasified about eighty thousand volumes of gaseous or ordinary air such as we breathe.

We illustrate the apparatus with which the experiments touching on this solubility of gases in liquid air were made at the Royal Institution by Dewar. *B* represents a cylindrical empty vessel of glass, something like a pipette in shape. It fits into a vacuum vessel, the joint between the opening of the vacuum vessel and the neck of the tube, *B*, being made tight by perforated stoppers. Through the central aperture of the cork or india rubber stopper, which is large, a branch tube passes, and through the center of this the neck of *B*, which is a capillary tube, passes. The whole is made air-tight by a perforated cork or india rubber stopper in the branch tube, through an aperture in which stopper this tube passes. A flask, *A*, contains liquid air, and a siphon, *H*, is so arranged that it delivers liquid air into the vacuum vessel, and keeps its level such that the tube, *B*, is constantly covered with liquid air. An air pump is connected above the neck of the vacuum vessel and keeps a high degree of exhaustion over the liquid air in *K*. The tube, *H*, from the flask, *A*, enters the vacuum vessel through the second aperture in the rubber stopper which closes the neck of the vessel in question.

The tube, *I*, leads to a gasholder full of air. This gasholder is graduated so that the air which it delivers is measured. Under the influence of the intense cold, air liquefies in the tube, *B*, coming from the gasholder and passing through the tubes, *C* and *D*, the lower one, *C*, charged with potassium hydrate, the upper one, *D*, with pumice stone saturated with sulphuric acid. Thus the air before it reaches *B* is thoroughly purified.

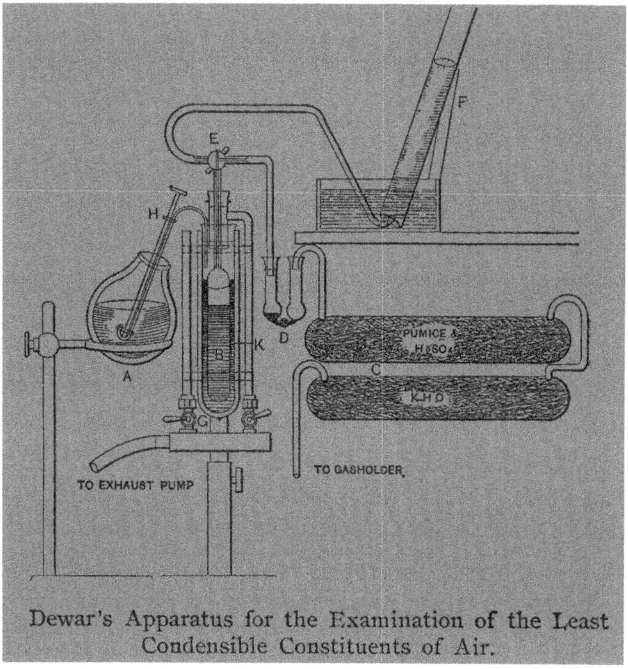

Dewar's Apparatus for the Examination of the Least Condensible Constituents of Air.

After forty minutes' operation with pure air the body of the tube, *B*, and the cool part of the capillary tube were filled with liquid, showing that everything delivered from the gasholder was liquefiable. From two and a half to three feet of air were used in each experiment. The capillary tube was so small and long that if only one volume out of 180,000 volumes of gaseous air had been unliquefied, it could have been detected. The first experiment showed complete condensation or liquefaction.

To the gasholder of 283 liters capacity (10 cubic feet) and holding that quantity of air, one-half a liter of hydrogen was added, which was in the proportion of less than one volume in five hundred. The experiment was repeated.

The tube, *B*, would not fill; only four-fifths of its volume was occupied by liquid, the other fifth was occupied by gas.

At *E* is a stopcock of the variety termed three-way. Turned in one direction, it connects *B* with *I*, *C*, and *D*, the air or gas supply. Turned in another direction, it connects *B* with the tube, *F*. Hitherto it had been turned so as to connect the air supply with *B*. Now it was turned so as to shut off the air and connect *B* with the tube, *F*. The temperature was allowed to rise a little, so that the gas from the upper portions of *B* bubbled up into *F*. The latter was originally filled with water. Its upper end, not visible within the limits of the cut, was closed.

The gas thus collected was tested and proved to be principally hydrogen.

Next air containing one volume of hydrogen in one thousand volumes of air was tried, and a very little hydrogen remained uncondensed. Finally, one volume of hydrogen was added to ten thousand volumes of air, and this liquefied completely.

Therefore, one volume of gaseous hydrogen in one thousand volumes of gaseous air can be almost completely liquefied. In the experiment, eighty thousand cubic centimeters of air were condensed to about one hundred cubic centimeters of liquid air, and dissolved eighty cubic centimeters of gaseous hydrogen. In other words, air liquefied at atmospheric pressure dissolves about eight-tenths of its liquid volume of gaseous hydrogen.

The apparatus just described was used for a most interesting piece of work, the separation of helium from the gas evolved from the water of the King's Well at Bath, England. This element, first discovered by spectroscopic observation in the sun and named from that fact, was not known to exist upon the earth. But some minerals were found to contain it in small quantities, and the gas from the Bath spring gave its spectrum. A good object for experiment was desired, which would show how applicable the method just described was for separation from each other of gases of varying degrees of ease of liquefaction.

The gas from the Bath spring contains a little over one-thousandth of its volume of helium (0·0012 vol.) The gasholder was filled with the gas, and the experiment just described was repeated. The tube, *B*, collected a liquid, not clear like liquid air, but turbid and yellowish brown. The color was found to be due to organic matter, probably of the petroleum family. Tested with nitric acid, it gave the familiar odor of nitro-benzole or of artificial oil of bitter almonds. This odor resembles that of the kernels of peach pits. It is sometimes used for perfuming soap.

After an hour some 20 cubic centimeters of gas had collected in *B* above the liquefied gas. Seventy liters of gas were liquefied.

The liquid in the tube was nitrogen. By letting the temperature rise, after properly turning the stop-cock, *E*, the gas along with some nitrogen was collected in the tube, *F*. The sample collected was about one-half nitrogen and one-half helium.

The experiment was extremely satisfactory as showing the practicability of using this liquefaction method for separating traces of less condensible gases from those which are more so. As Prof. Dewar observes, a regular gas liquefaction apparatus could be installed at Bath and made to produce any quantity of helium, were there any demand for it.

In this class of experiment we see fractional condensation, long since applied in distillatory processes, applied to gases. It is an interesting subjection of the most elusive substances to processes hitherto only applied to ordinary liquids.

A rather interesting demonstration of the action of mixed gases when liquefied in presence of each other was afforded by the liquefaction of oxygen in the presence of an excess of hydrogen. The liquid, as we have seen, could contain but little hydrogen. Yet the gas given off by it contained so much that it was explosive. In the evaporation, naturally a much larger relative proportion of hydrogen evaporated than of oxygen, so that the gas contained perhaps over one-half its volume of hydrogen, while the liquid, as we have seen, could contain but a little more than a trace dissolved.

One of the recent triumphs of chemistry was the isolation of fluorine. For generations of chemists it had proved an element which could not be separated from its compounds. It has most intense affinities for other elements, and attacks glass with much energy. Moissan, a French chemist, succeeded in separating it in the elemental state. In 1897 Moissan and Dewar, working together, liquefied it.

From theoretical considerations it appeared that fluorine should be more difficultly liquefiable than chlorine. Thus boron fluoride and silicon fluoride are gases, the corresponding chlorides are liquids. The same holds with many organic compounds—those containing chlorine being liquid and those containing fluorine being gaseous. This, obviously enough, was taken as indicating that fluorine was more difficult to liquefy than chlorine.

The experimenters made fluorine by electrolyzing a solution of potassium fluoride in hydrofluoric acid. The gaseous fluorine evolved was passed through a platinum condenser tube which was cooled by solid carbon dioxide mixed with ether. This was intended to condense all impurities. It passed through another platinum vessel filled with perfectly dry sodium fluoride and then into the liquefaction vessel.

One of the great troubles of fluorine, as a subject for experiment, is that it attacks glass. For this reason platinum vessels are used for accurate work with it and its compounds. Lead stills and flasks are used for rough work, and the natural mineral fluorspar has even been suggested as a material for vessels.

The liquefaction vessel was a glass capsule into whose upper part a platinum tube was soldered.

The tube from the fluorine evolution and purification apparatus entered the outer tube and passed down the annular space into the glass cylinder or capsule. The latter was immersed in liquid oxygen, which, boiling at atmospheric pressure, gave a temperature of —183° C. (—297·4° F.) The glass was not attacked at this low temperature, and the fluorine did not liquefy. Exhaustion was now applied to the oxygen, and the reduction of pressure reduced the temperature to about —187° C. (—304·6° F.) A dew of liquefied fluorine began to appear upon the glass.

In the first experiments the platinum tube leading out of the vessel had no cock. Upon closing it with the finger, fluorine at once began to collect in the glass capsule, which rapidly became partly filled with it. It was a clear, very mobile liquid of yellow color. The intensity of the color was stated to be equal to that which would be given by a column of gaseous fluorine one meter long.

The liquid was so cold as to have little chemical power left. A number of substances were tried. Silicon, boron, carbon, sulphur, phosphorus and iron reduced in hydrogen could, after cooling with liquid oxygen, be dropped into it without any reaction. Ordinarily, fluorine would attack them violently. At the temperature of —180° C. (—292° F.) it attacked benzene and turpentine. It could not separate iodine from potassium iodide. Hydrogen burned upon the surface of the liquid when caused to impinge thereon.

It was cooled to —210° C. (—346° F.) by boiling liquid air, in hopes that it would solidify, but it remained liquid. By accident, some air got into the capsule of liquid fluorine. It liquefied and floated upon it, a colorless or faint blue liquid upon the pale yellow fluorine. But, by passing a current of fluorine through liquid air, a flocculent precipitate formed. This was filtered out, and on heating exploded with great violence. In a subsequent experiment the same layer of fluorine under the liquid oxygen just described was formed by passing fluorine to the bottom of a vessel of liquid oxygen. Evidences were found that liquid oxygen would dissolve it under certain conditions, the fluorine being admitted, not to the bottom, but to the surface of liquid oxygen. The subject remains obscure.

The specific gravity was determined by placing in it different substances of known specific gravity and observing which ones floated and which ones sank. Ebonite, caoutchouc, wood, amber and methyl oxalate were taken. The pieces were placed in the empty tube, and fluorine was liquefied in it. Wood, caoutchouc and ebonite floated, the methyl oxalate sank, and amber was almost indifferent. This gave it the same specific gravity approximately as that of amber, or 1·14.

The amber could only be seen with difficulty, so that the refractive index of liquid fluorine is almost the same as that of amber.

On cooling it from —187° C. (—304·6° F.) to —210° C. (—346° F.), it diminished one-eleventh in volume. It possessed no magnetic features as far as tested.

Its capillarity is less than that of liquid oxygen. The relative heights to which it and other liquids rise in a capillary tube were determined, with the following results:

Liquid fluorine..	35	Alcohol.........	140
Liquid oxygen...	50	Water...........	220

Water, therefore, rose about seven times as high as fluorine.

May 10, 1898, is one of the classic dates in our subject, for it was on this day that Dewar liquefied hydrogen, and obtained it in quantity as a "static liquid."

A very powerful train of liquefying apparatus had been set up in the Royal Institution, its erection extending over a year's time. It weighed two tons and contained 30,000 feet of piping.

Hydrogen was cooled to —205° C. (—337° F.) at a pressure of one hundred and eighty atmospheres. The gas was allowed to escape continuously from the nozzle of a coil of pipe, at the rate of ten or fifteen cubic feet a minute. When it is stated that an ordinary gas burner burns about six cubic feet per hour, it will be seen that hydrogen was used most profusely. The jet issued into a doubly silvered vacuum vessel, surrounded by another vessel, the intervening space being kept at —200° C. (—328° F.) Soon drops of hydrogen began to appear, and in five minutes twenty cubic centimeters had collected. The goal was won. Static hydrogen lay quietly in a vessel.

The jet then closed with frozen impurities from the hydrogen. One per cent. of the gas had been collected in the liquid form.

A small glass bulb was weighed in the liquid and gave a specific gravity of 0·08—an amazingly low figure. The end of a long glass tube sealed at the bottom was placed in it, and at once became filled with solid air. Liquid oxygen was placed in a tube and immersed in it, when a blue solid was produced from the frozen liquid. It was solid oxygen, or oxygen ice.

A glass tube closed at its upper end was placed in a vertical position with its lower open end immersed in a vessel of mercury. It was so arranged that its upper end could be cooled by liquid hydrogen. On doing so, the mercury rose in the tube as the air solidified, until it stood within a minute fraction of an inch of the height of the barometric column.

If liquid hydrogen were placed in a double-walled non-exhausted vessel, it froze the air in the interspace solid, and the inner vessel became coated with a hoar frost or coating of solid air, literally of air-ice. The liquid hydrogen manufactured its own Dewar's bulb.

A metal rod dipped in it became so cold that, on removal, liquid air fell from it in drops, liquefied by the cold of the rod due to its immersion in the liquid hydrogen.

A sample of the helium obtained by Dewar from the gas of the Bath spring (page 275) was at hand in a sealed bulb with a narrow tube attached to it. The tube was dipped into the liquid hydrogen. Liquid helium formed in it as a distinctly visible liquid.

As a control experiment, the same tube was put into boiling air and no liquid formed. This showed that the cold of boiling air was insufficient to produce a liquid from it; the liquid hydrogen gave a degree of cold sufficient to do it.

The boiling point of the liquid hydrogen in the first experiments was determined by a platinum resistance thermometer. At 0° C. (32° F.) this had a resistance of 5·3 ohms. In the liquid hydrogen the resistance fell to 0·1 ohm. From the observation the temperatures of —238·2° C. (—396·76° F.), —238·9° C. (—398° F.) and —237° C. (—394·6° F.) were calculated on slightly differing bases. These temperatures are about 8° C. (14·4° F.) higher than Wroblewski's calculated temperature of boiling hydrogen, and 5° C. (9° F.) higher than that given by Olszewski's calculation.

In later experiments the following results were obtained: The resistance of the platinum wire resistance thermometer sank from 5·338 ohms at 0° C. (32° F.) to 0·129 ohm at the boiling point of hydrogen. This gave the boiling point as —238° C. (—396·4° F.) The resistance of the platinum wire in liquid oxygen was eleven times that of its resistance in liquid hydrogen, both at atmospheric pressure. At its boiling point the pressure of air, which is solid at that temperature, is but 0·002 millimeter of mercury. This is one three hundred and eighty thousandth of the normal pressure. The vapor density of hydrogen at the temperature of its boiling point is eight times greater than at ordinary temperatures, or about one-half as heavy as air at ordinary temperatures.

The critical temperature is about 50° C. absolute (90° F. absolute) and the critical pressure is less than fifteen atmospheres. The latent heat is about two-fifths that of oxygen. The application of a vacuum to liquid hydrogen, therefore, cannot lower its temperature very much, compared with the cases of other gases.

An approximate determination of the density was made by measuring off ten cubic centimeters of the liquid, and collecting and measuring the hydrogen gas from it. The result was 0·07—not far from that obtained by weighing the glass bulb in it. It is about one-sixth that of liquefied marsh gas (0·41).

The light, evanescent liquid is, nevertheless, perfectly visible, has a defined meniscus, and can be readily manipulated in vacuum vessels.

The atomic volume at the temperature of its ebullition is 14 3 (oxygen=13·7; nitrogen=16·6). The gaseous hydrogen at this temperature has a specific gravity of 0·55 (air=1). The ratio of the specific gravity of the gas, compared to that of the liquid at the ebullition point, is as 1 : 100 (oxygen=1 : 255).

The specific heat of gaseous hydrogen and of hydrogen occluded in palladium is 3·4; of liquid hydrogen, 6·4. The specific heat of the liquid, per unit volume, is 0·5, or about that of liquid air.

Liquid hydrogen affords a rapid means of obtaining one of the nearest approaches to a perfect vacuum which man can produce. We have just seen that air is solidified by the cold of liquid hydrogen. A tube is filled with air and sealed. The end of the tube is placed in liquid hydrogen. With surprising rapidity the air in the tube solidifies and collects in the lower end where immersed in the liquid, and a vacuum, almost perfect, is formed in the rest of the tube. An immersion of one minute in never exceeded. The tube, while its end is still immersed, is softened with the blowpipe flame above the hydrogen vessel, or as near where it emerges therefrom as possible, and under the effect of atmospheric pressure it closes and is sealed off. Thus a vacuum tube is produced without pump or other apparatus of similar function. The process is so simple and efficacious that it would seem to give a suggestion for the production of other vacuous vessels, such as incandescent lamps. A more easily solidified gas could be substituted for air, and liquid air could take the place of hydrogen. Sir William Crookes, celebrated for his work on high vacua, from whom the vacuum tubes used in high vacua experiments are named, examined these tubes. He found that a higher vacuum was produced than he was in the habit of getting in his own tubes, after several hours' work with the mercury pump.

On spectroscopic examination, the spectrum of carbon and of hydrogen was obtained. Neon and helium lines were also found. The carbon spectrum is attributed to carbonates in the glass.

An actual trial was made to determine what lowering of temperature would result from reducing the pressure under which the hydrogen boiled. As has been already stated, no great reduction was anticipated; not over 9° C. (16·2° F.) Under an exhaustion of one inch of mercury, very little lowering was effected. The extent of reduction due to the partial vacuum only amounted to 1° C. (1·8° F.) Possibly the platinum thermometer did not give the right result; possibly the connections conducted heat; possibly the resistance curve of platinum cannot be relied on at such excessively low temperatures.

With the liquefaction of hydrogen in bulk the story of the liquefaction of gases culminates. The date is but a few months before the period in which this book was written. It seems a most appropriate time in which to put together the long chronicle of a hundred years' efforts to liquefy gases, and whose final triumphs are no less Tripler's great buckets of liquid air, made in the city of New York, and sent off hundreds of miles by rail, than they are the few teaspoonfuls of liquid hydrogen liquefied by Dewar and his colleagues in the Royal Institution in London.

Hydrogen has been treated as a metal. In its liquefaction many expected that a metallic liquid like mercury would result. But the product was not in the least metallic, and was a non-conductor of electricity, so that a much mooted question as to the nature of hydrogen is at last settled.

CHAPTER XII.

Charles E. Tripler.

The life of Charles E. Tripler—His early experiments with gas motors—Mechanical difficulties encountered—His electrical experiments—Chemistry—His work in fine art—Exhibition of his paintings—Return to the investigation of compressed gases—Liquefaction of air—He endeavors to utilize the low grade heat of the universe—Simplicity of his apparatus—The plant—The compressor—General plan of operations—Capacity of his plant—How he transports liquid air—His lectures—Raoul Pictet in Charles E. Tripler's laboratory.

Charles E. Tripler was born in New York, August 10, 1849. From his early years he showed a great fondness for mechanics and experimenting, which fondness soon developed into practical work. In the early seventies his attention was directed toward the production of a motor to be driven by gas. He experimented on an engine driven by ammonia. His work was different from that of others in one important respect. The endeavor had been to actuate an engine by the pressure of ammoniacal gas, and to reduce its pressure by dissolving it in water.

This process Tripler wished to avoid. He desired to work the ammoniacal gas in a continuous cycle without having resource to solution. Gasolene and naphtha were next tried, much trouble being experienced in those early days with the joints in the apparatus, high pressure work in engineering having greatly developed during the last twenty-five years. One of the objects was to produce a motor for use on street cars.

Electricity and chemistry were now (1873–76) taken up. Edison was at the same time engaged on electrical problems, and Tripler left the field and took up art.

An artist by nature, he painted and exhibited paintings, and left his mechanical and scientific work almost untouched for a few years.

About 1884 he worked on gold extraction and amalgamation and then returned to his first love and experimented with gases of many kinds, ethyl chloride, methyl chloride, and at last with carbon dioxide. During these researches he discovered the principle on which his work on the liquefaction of air has been based.

Nitrous oxide was the next gas to be experimented with, and an explosion brought about during the generation of the gas nearly cost the investigator his life. His work, being at high pressure, and with many gases, has always been attended with peril, and the wholesale manipulation of liquid air is far from safe, irrespective of the question of pressure and danger of explosion. All sorts of gases were made and liquefied, and about 1891 air was liquefied.

The key to his life's work has been the effort to use gases for motive power, Carnot's cycle giving the clew to what he has desired to accomplish.

He desired to utilize the heat of the sun. If the first chapters of this book have been followed out to their conclusions, it will be seen that the utilization of the low grade heat energy of the universe, in accordance with Clerk Maxwell's dream, presents nothing of the essentially impossible. This heat Tripler hopes to utilize. If it is utilized, there will be a further demand made upon the heat of the terrestrial system, which will involve a reduction of temperature due to the conversion of low grade heat energy into mechanical energy. This involves a theoretical loss of temperature by the earth and its atmosphere from self-contained causes, and the loss would have to be replaced by heat derived from the sun.

Perhaps the most striking feature about the Tripler process, apparatus and plant is that there is comparatively little to be said about it. While Dewar, working on the lines laid down years before by Pictet and assisted by liberal gifts from one of the London guilds and from private individuals, liquefied gases at vast expense, here in the metropolis of this country a private individual has erected a plant at his own expense, and for years past has liquefied air on a scale which Dewar and his associates never dreamed of. In order to preserve air, Dewar devised his celebrated vacuum bulb, an apparatus of the highest merit. Tripler took common tin cans, lined them with felt, filled them with two to five or more gallons of liquid air, and sent them off hundreds of miles by rail.

In the reports of papers and discussions in English scientific gatherings incredulity is still expressed, or was until very recently, when the sending of liquid air about in common tin buckets was spoken of.

In England, Dewar has excited the greatest enthusiasm by his lectures on liquid air and liquefied gases.

The enthusiasm was deserved, and it is a hopeful sign of the times that a popular audience can still be so stirred to a high pitch of interest in a scientific subject. But, meanwhile, Charles E. Tripler, in his private laboratory, with boiler, air compressor and simple liquefying apparatus, has repeatedly shown liquid air, in quantities that until recently scientists would scarcely have believed possible of production, has poured it out on the floor by gallons to show its rapid evaporation and production of dense clouds of condensed moisture, has blown iron pipes to pieces with it, and has permitted physicians to try its effects as a cautery upon patients.

Pouring out Liquid Air on the Floor in Tripler's Laboratory.

Mr. Tripler's apparatus is of the type which employs no extraneous sources of cold. All the liquefaction is done by its own powers and within its own system. A steam boiler is installed in one corner of the laboratory in which his plant has been erected. This supplies steam to a Norwalk straight line compressor. The steam pressure is about 85 pounds to the square inch.

The compressor is a steam engine with three com-

Tripler's Laboratory, Showing Air Compressor and End of Liquefier.

pression cylinders in line of the prolongation of the axis of the cylinder. The piston rods run in one line through the four cylinders. The engine is rated at 90 horse power when working at 150 revolutions. For the work done in Mr. Tripler's laboratory the rate is about 100 revolutions.

The stroke of the engine, and, consequently, that of the four compression pistons, is 16 inches. The steam cylinder is of 16 inches diameter, the first or low pressure air cylinder is of 10½ inches diameter, the intermediate cylinder is of 6⅝ inches diameter, the high pressure, the last of the three, is of 2⅝ inches diameter. The pressure is brought up by three steps. The first compression raises it to a pressure ranging from 55 to 65 pounds above the atmospheric pressure; the next compression, from 350 to 400 pounds; and the final from 2,000 to 2,500 pounds per square inch. The areas of the pistons in the three air compressing cylinders are in the ratio of 110 : 44 : 6, the air pressures successively produced as 55 : 350 : 2,500.

The cut gives a diagrammatic representation of the general arrangement of the apparatus in Tripler's laboratory, and the cut on page 291 gives a view of the interior. On the left is seen the boiler, and in the background is the compressor. The three air cylinders of the compressor are arranged in tandem or in line with each other. Between the first and second and between the second and third air cylinders are surface condensers which cool the air. Compression, as has been explained, heats a gas.

The air is drawn down from the roof of the building through a pipe, and goes through a washer

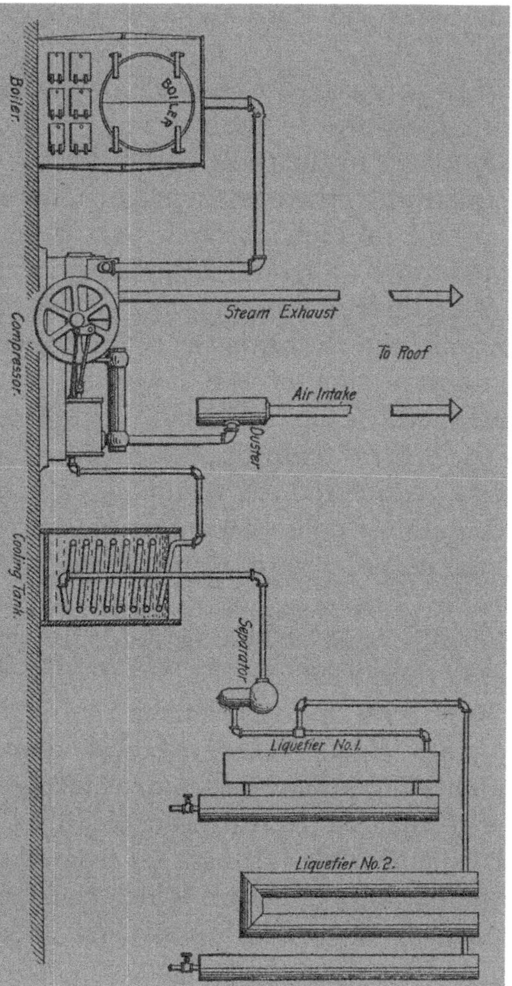

Diagram of Arrangement of Tripler's L q il Air Plant

which removes the dust. This is a case containing baffle plates over which water is kept trickling. It is marked "duster" in the diagrammatic cut. The air then goes through the compressor with its coolers and leaves the third cylinder at high pressure and hot.

The heat is removed by a final cooling in a surface condenser designated "cooling tank" in the diagram.

The moisture in the cooled air is pretty thoroughly precipitated by the compression. There are some traces of oil present, derived from the lubricating oil of the pump. Such of this material as is carried forward is removed in a separator, which is virtually a steam-trap, and the air is ready for liquefaction.

The construction of the liquefiers has not been fully divulged. The lower end of one is seen in the cut on page 291. They appear as long felt-covered cylinders. Inside the felt wrappings are cylindrical cases containing coils of copper pipe. At the bottom of the coil of pipe is a special valve, the invention of Mr. Tripler. The compressed air escapes from the valve and, expanding suddenly, experiences a drop in temperature. Some of the cooled air works its way up through the chamber and cools the coils of pipe. Thus there is established an intensive or accumulating action. The air entering the liquefier at a normal temperature is cooled by the reverse flow of expanded air. It escapes from the valve at the bottom at a temperature which constantly grows lower until air begins to liquefy, and collects in the bottom of the liquefying chamber. Now all is in working order, air is liquefying and collecting, and in a short time liquid air can be drawn off by the gallon just like water.

Three or four gallons of liquid air are produced in an hour in the usual operation of the plant, but power enough is present to produce far more. Every part of the liquefiers is insulated with non-conducting covering. Only the handles of the valves protrude, and these become white with a thick deposit of hoar frost.

The diagrammatic cut gives a general idea of the distribution of parts, but is not given as a representation of the plant in any sense.

One of the most remarkable things about Mr. Tripler's work is its simplicity even in detail. There is no refrigerant used, and nothing is to be seen but the ordinary objects which meet the eye in any steam plant. There are no cylinders of liquefied ethylene or carbon dioxide. Even the compressor is of a normal type. Yet in this apartment the most impressive achievement in physics of the century is repeated week after week, and air is liquefied by the bucketful and handled as if it were so much water.

Its transportation is interesting. No vacuum bulbs are needed for this. A tin bucket is wrapped with boiler felt and is thrust into a larger one. The liquid air is poured into the inner bucket, a piece of felt is placed over the mouth, and the air is ready for removal. In such buckets it has been taken hundreds of miles.

In the cut on the next page are given sections of two of the buckets, one holding twice as much as the other. The scale is 1½ inches to the foot.

Mr. Tripler has given many lectures on the subject of liquid air, and in the next chapter are illustrated a number of the experiments which he shows. But his goal is the practical, and his lectures are merely a side issue and express only his deep interest in the subject.

An interesting occasion was the presence of Prof. Raoul Pictet at one of Mr. Tripler's demonstrations. The American inventor tells of Pictet's enthusiasm at witnessing the demonstrations executed with such

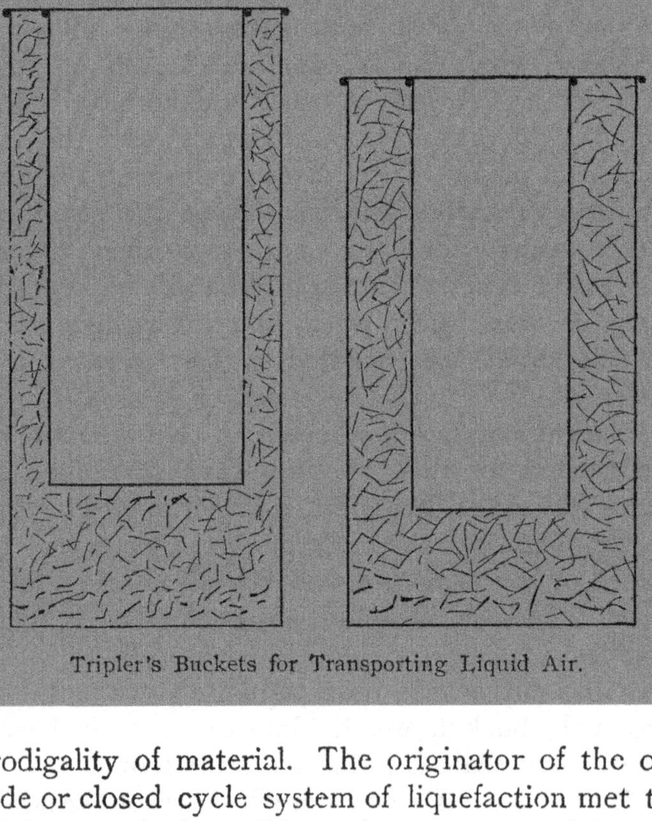

Tripler's Buckets for Transporting Liquid Air.

prodigality of material. The originator of the cascade or closed cycle system of liquefaction met the originator of the self-intensive system only to be delighted at his demonstrations.

CHAPTER XIII.

THE JOULE-THOMSON EFFECT.

First attempts at liquefying gas—Joule and Thomson and their discovery—Coal a cheap chemical—Substitution of mechanical for chemical energy—Sir William Siemens' regeneration of cold—Self-intensive refrigeration—Negative Joule-Thomson effect—Mathematics of the theory—Conditions of pressure for economical application.

The first attempts at liquefying gases were based on the application of great pressure. This was at once useless and unnecessary in many cases; useless because an insufficient lowering of temperature was applied and the gases did not liquefy, and unnecessary because the high pressure was not needed, had a sufficient refrigeration been applied. Cailletet, and probably Pictet, got useful effects indirectly from high pressures. By sudden release of high pressure a great refrigeration was produced, the temperature of the gas fell below the critical point and it liquefied.

The discoveries due to Joule and Thomson that air and most gases are not perfect gases, that there is really no perfect gas, and that hydrogen is an ultra-perfect gas, has already been spoken of on pages 60 *et seq.* The change of temperature in a given mass or volume of gas brought about by letting it flow under pressure through an orifice, an effect not to be confused with cooling due to expansion, while so trifling as to have entirely escaped recognition by Joule in his early experiments, has been taken as the starting point for the operation of refrigerating machines. The movement, whether we accept the theory of action offered or not, was in the direction of purely mechanical production of cold, and hence was in the direction of economy. Dewar speaks often of the great expense of his liquefactions, in effecting which a very large expenditure was incurred in the production of liquid ethylene alone, so that the cost of this and of other refrigerants was a large item of expense in the Royal Institution work.

In general terms we may say that coal is the cheapest chemical we possess. Could the old time experimenters have seen the possibility of substituting coal for the expensive liquefied ethylene and other gases, they would have been most delighted. In the processes of liquefying air and oxygen which we are now to describe this in a sense is done. Air is liquefied by the application of power, and neither liquid ethylene, solid carbon dioxide nor other refrigerant is needed. Even coal may be dispensed with, for the energy of a waterfall might be utilized to produce liquid air.

As a general rule, it may be stated that the substitution of mechanical power for chemical and for other special agents is one of the most impressive movements of the age. The electric battery giving way to the mechanically impelled dynamo is an excellent example of the movement alluded to. In the field of refrigeration the substitution of a purely mechanical process for refrigeration by boiling liquefied gases was to be greatly desired, and in the application of the Joule-Thomson effect the possibility has been claimed of effecting the substitution.

When gas expands under terrestrial conditions, it practically always falls in temperature. It is not easy to see how conditions could be established which would expand a gas without such fall. This fact was well known for many years, and over forty years ago the idea of applying it to refrigeration and of making it more effective by cold regeneration was suggested. It was William Siemens who saw the possibility of utilizing it by a regenerative process for the production of still lower temperatures. It is fair to presume that his mind was, at the period in question (1857), deeply occupied with the subject of the regeneration of heat, and the regeneration of cold seemed a natural sequence of the other. He simply thought of the cold due to the energy developed by an expanding gas. This development of energy calls upon an equivalent quantity of energy for its development, and in the case of an expanding gas the energy which is called upon is the heat energy of its molecules. This heat energy is converted into mechanically exerted energy and disappears as heat—therefore cold is produced.

Leaving out of account this refrigeration, we know that, if a gas is expanded, there is another change in temperature outside of and independent of the natural cooling due to energy developed in expansion. This is what we have termed the Joule-Thomson effect. The apparently slight refrigeration thus produced is the principle claimed to underlie the operation of two of the most prominent of the gas liquefaction processes now in use. Linde's and Hampson's apparatus are the ones alluded to.

There is nothing of efficiency involved in the small orifices or porous diaphragm as used in the experiment. It is simply a way of localizing expansion and of producing it. As it is an element of the most practicable way of rendering possible the expansion of a gas from a high degree of compression, it is always used, but there is nothing occult about it as far as the valve or aperture is concerned, outside of mechanical advantages.

The term self-intensive refrigeration is perhaps etymologically preferable to regeneration. This preference would be based on the idea that the production of cold is not, properly speaking, an operation involving production, but destruction. Cold is the negation of heat, and, properly speaking, cannot be said to have an existence of its own. But William Siemens, doubtless thinking over his methods of regenerating heat, in his 1857 patent prescribes the regeneration of cold.

The origin of the methods used by Tripler, Hampson and Linde can be studied in the records of the patent offices as well as in the literature of pure science.

The primary idea of the self-intensive process is found in the Siemens provisional specification of the English Patent Office. He simply contemplates cooling air by expansion, thereby causing its heat energy to disappear. This cooler air is caused to act upon the entering air, and give it a lower temperature before expansion, so that the cold grows constantly more intense. But Siemens has no idea of utilizing the expansion through a small orifice, which is the system so much employed at present. The Joule-Thomson effect was not known at the early date which we speak of.

In 1893 Tripler applied for and was granted a patent by the English Patent Office for a gas liquefying process and apparatus. This most interesting document gives a clear description with drawings of an apparatus based on self-intensification for the production of cold. The Joule-Thomson effect is not appealed to in it.

It is far from certain that the Joule-Thomson effect is the principal factor in the operation of modern self-intensive gas-liquefying machines, even if we admit Onnes' theory that the negative effect which obtains with hydrogen is reversed under more extreme conditions. We are justified in attributing especial importance to such utterances as those contained in Siemens' early provisional specification, and in Tripler's early patent, which is full and complete and is illustrated by drawings.

The use of an aperture for expanding gas through is more justly regarded as an expedient for readily bringing about a great difference in pressure in a gas or, what is the same thing, for causing a great expansion and sharply locating it.

But whatever influence the Joule-Thomson effect has, whether great or small, Linde and Hampson have both invoked it as the principle on which their machines operate. It is easily stated, and involves in its study but little mathematics. In Cailletet's and in Wroblewski and Olszewski's liquefactions by release there was no thought of appealing to this almost trifling effect to account for the mists of oxygen and other gases observed when they suddenly expanded. The cloud of moisture seen in the receiver of a common air pump with the first strokes of the pump were never supposed to be due to it. It is not clear why it has to be invoked as the factor in liquefying air by the gallon.

The theory may be thus stated:

If air be expanded through a fine orifice, the change in temperature due to the Joule-Thomson effect is thus calculated:

$$\text{Fall in temperature} = \frac{p^2 - p^1}{4}\left(\frac{289}{T^1}\right)^2$$

In this formula p^2 is the pressure in atmospheres before passage through the orifice or before expansion, p^1 is the pressure after passing through it or after expansion, T^1 is the temperature of the gas before passing through it in degrees Centigrade referred to the absolute zero.

The work which a pump has to do in forcing a continuous circuit of air round and round through this aperture varies with $\frac{p^2}{p^1}$ This is because the work of the pump depends on the ratio of pressures on the front and back of the piston. The greater the pressure in front in proportion to the pressure back of it, the more work it has to do.

To get a good reduction of temperature, it is evident that the quantity $p^2 - p^1$ of the first formula must be as large as possible and the quantity T^1 of the same formula must be as small as possible. The first of these is regulated by the proportions given the different parts of the apparatus, the second quantity grows smaller as the temperature of the gas to be liquefied falls. In circulating apparatus, this temperature, as we shall see, falls continuously, the longer the apparatus is worked, until air begins to liquefy.

The ratio $\frac{p^2}{p^1}$ may be kept small and the difference $p^2 - p^1$ large by giving high values to p^2 and p^1; in other words, by working at high pressures.

A formula often seems uninteresting, but if the substitutions of real values for the letters are made, it acquires concrete interest.

Assume that the air, in passing through the orifice, falls 3·6 atmospheres in pressure, and assume that we start with a temperature $T^1 = 0°$ C. $= 273°$ C. absolute. The fact that the fall in pressure is 3·6 atmospheres makes $p^2 - p^1 = 3·6$. Our formula now reads:

Fall of temperature $= \frac{3·6}{4} (\frac{289}{273})^2 = 1°$ C. (1·8° F.)

This seems a very trifling fall of temperature.

But assume that the air is driven more vigorously through the orifice until a difference of pressure of ten atmospheres is maintained, then the formula reads:

Fall of temperature $= \frac{10}{4} (\frac{289}{273})^2 = 2·78°$ C. (5° F.) which is at least somewhat more appreciable. So it follows that by changing the mechanical relations we can produce falls of temperature of various degrees.

On inspection of the formula another thing becomes evident. The lower the temperature before passing the orifice is, the greater will be the fall in temperature. To assume T^1 to be —91° C. (—131·8° F.), which is in absolute degrees C. 273—91=182° C., the quantity $\left(\frac{289}{T^1}\right)^2$ reduces to the factor 2·52 in round numbers; so that if the gas, as it reaches the diaphragm, can be got down to this temperature, the fall in temperature will be greater in the ratio of $(\frac{289}{273})^2 : (\frac{289}{182})^2 = 1{\cdot}12 : 2{\cdot}52$, or 1 : 2·25, also in round numbers. Hence, at this temperature, for the two pressure differences we have taken, the temperatures would be 1° × 2·25 = 2·25° C. (4·05° F.), and 2·78° × 2·25 = 6·26° C. (11·26° F.)

The first substituted formula has been purposely constructed so as to give a temperature fall in round numbers of 1° C. If there is a different pressure drop employed, the fall of temperature due thereto when $T^1 = 273°$ C. absolute or 0° C. is found by dividing the pressure drop expressed in atmospheres by 3·6 and multiplying by unity. This gives directly the fall in temperature.

Thus, if a fall of 10 atmospheres were to be assumed, we have 10 ÷ 3·6 = 2·78, which, multiplied by unity, gives 2·78° C., as calculated by the second substituted formula.

Assume now that we are working with a different temperature, T^1. Then we may divide it by 273 and square the product and divide unity with it, and the result will give the degrees Centigrade of fall of temperature at a pressure drop of 3·6 atmospheres. Thus suppose the temperature T^1 to be —91° C.

This is 182° C. absolute. $\frac{182}{273} = \frac{2}{3}$, which squared is $\frac{4}{9}$. To divide unity with it, we invert and multiply, which is expressed thus: $\frac{9}{4} \times 1 = 2{\cdot}25$. This is the factor used in the third substitution example.

It is evident that with a formula for a fall of temperature $= \frac{3{\cdot}6}{4} \left(\frac{289}{273}\right)^2 = 1°$, we can, by applying thereto the two methods of calculation last described, make it apply to any case. Thus, if we assume that the pressure drop is 10 atmospheres and that $T^1 = -91°$ C., we have simply to multiply unity by one of the factors already determined, and this product must be multiplied by the other. These factors are 2·78 and 2·25; we have, therefore:

$$1 \times 2{\cdot}78 \times 2{\cdot}25 = 6{\cdot}255° \text{ C. } (11{\cdot}27° \text{ F.})$$

The same result could be reached by substituting directly in the equation

$$\text{Fall of temperature} = \frac{p^2 - p^1}{4} \left(\frac{289}{273}\right)^2$$

These examples merely illustrate different ways of reaching the same results.

The statement has been made that the power required to force air through the aperture varies with $\frac{p^2}{p^1}$ in which p^2 is the pressure in the inlet side of the aperture and p^1 the pressure of the gas after it has passed through it. The reason of this proportion existing is due to the fact that gas is diminished in volume by pressure. Thus, if a given weight of air is to be pumped through an aperture by a pump, it may be done at very low pressure or at high pressure. At first sight it might be thought that at high pressure, when the pump is working against a pressure of fifty pounds to the square inch, more power would be required than when it works against a lower pressure. But, air being compressible, the pump at high pressure has a less volume of air to force through, and hence has fewer strokes to make.

The air which enters the suction end of the pump may be looked upon as reinforcing its action. Hence the higher its pressure is, the less work will the pumps have to do. Hence the smaller p^2 is and the larger p^1 is, the less work will the pump have to do.

CHAPTER XIV.

THE LINDE APPARATUS.

Linde's apparatus—The simplest form of apparatus—Its operation—Its storing of air at atmospheric pressure—Avoidance of atomization and waste—Subdivision of pressure-drop—Laboratory apparatus—A feature of inefficiency in it—Its power of liquefaction—Continuous oxygen-producing apparatus—Date of Linde's first successful use of his apparatus.

Linde's apparatus, which is described as utilizing this small increment of cold, if the expression may be allowed, and by constant summation of such increments bringing about a high degree of refrigeration, caused much interest when its supposed principles were first stated and its operations were first disclosed. The term self-intensive has been aptly coined to describe machines of this type.

What the apparatus of the original Linde type does is this: Air is pumped through a circuit of pipes; the pipe from the outlet of the pump, after going through the given circuit, returns to the inlet, so that the air under treatment goes constantly around the same circuit. When a gas is pumped against resistance, it is compressed or diminished in volume and heated. The outlet pipe from the pump is kept at a uniform temperature by cold water circulating in contact with the outside of the pipe, like a surface condenser.

The air thus cooled is forced through a small aperture, and the passage from high to low pressure, with consequent expansion, causes cooling. Between the water cooling apparatus and the aperture a long length of pipe intervenes. The cooled air is carried back to the pump so as to circulate around this pipe on its way back, and it abstracts heat from the air already cooled by the water. Hence the air reaches the aperture constantly at a lower temperature, but leaves the water condenser always at a uniform temperature. The real cold production is done after the air leaves the water condenser. The degree of cold keeps increasing until liquid air drops from the aperture and lies in the bottom of the apparatus. By a cock it can be drawn therefrom like water.

It seems at first sight impossible that the small decrease of temperature, due to the imperfection of the gaseous state as it exists in air, should be able to produce such refrigeration. What Hampson calls thermal advantages are to be aimed at. The surface on which the cooled air acts on its return must be large, the material of the pipes thin. These elements provide for a rapid cooling by the returning air of the counter-stream on its way to the aperture. The entire mass to be cooled must be as light as possible. The action of the pump is constantly heating the gas by compression, and this heat is removed by the water. The atmosphere surrounding the apparatus constantly heats the portions colder than itself by contact. The colder portions, therefore, must be protected from this action by thick jacketing or other means. Concentric air spaces produce a good effect, and doubtless if it were practicable Dewar's vacuum heat insulation might be applied with excellent effect.

Linde made quite a sensation by his description of his apparatus, which, by purely mechanical means, liquefied air, although his first results were far from encouraging.

What is called Linde's simplest form of apparatus is illustrated in the cut, and will be readily understood, especially if the reader has grasped the very simple general theory on which its operation depends. It will be understood that the drawing is not a reproduction of the exact apparatus, but is diagrammati., being purposely made as clear as possible without permitting detail to interfere with intelligibility.

P represents a pump which aspirates air from the pipe, *G*, and forces it out, under pressure, through *H*. The air forced out through *H* goes through a complete circuit of pipes and returns through *G*, thus constantly and repeatedly going around the circuit.

J is a water condenser or more properly a cooling apparatus. It is a cylindrical vessel, and the air pipe goes through it in a coil. Water enters at *K* and emerges at *L*, so that as the gas leaves the vessel it is always at the temperature of the inflowing water. The arrows show the direction of the current of gas, and all is perfectly clear to the point, *C*. The arrows might be taken to indicate that the gas, on reaching *C*, goes down directly to *G*, but they do not indicate this. The pipe, *B*, is of small diameter, and, without any opening or break, runs straight on to *D*, is bent into a coil, and descends to *E* and *T*. But from *C* to *F* it is surrounded by a second pipe concentric with it, and it is this outer pipe which is connected to the pump suction by the vertical pipe extending downward from *C* and ending in *G*.

The cylindrical vessel on the right is simply a non-conducting casing or jacket to protect the pipes from the heating effect of the outer air. In the illustration

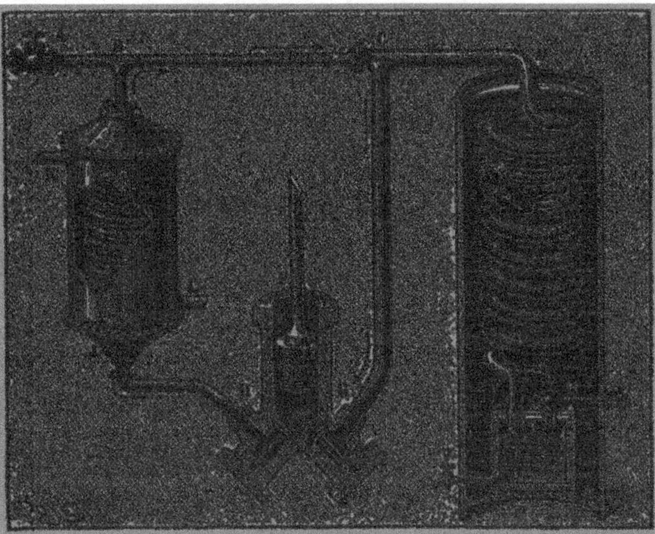

Linde's Apparatus for Liquefying Air.

the interior of the coil is shown, a part of the pipe being supposed to be broken away to show this.

In the course of the air in the pipes to the right of the point, *C*, lies the soul of the apparatus. The small pipe running down through the protecting vessel terminates in the chamber, *T*. A valve, *R*, is provided which may be opened or shut so as to regulate the pressure drop, and this valve constitutes the aperture through which the gas passes and expands with attendant cooling.

The end of the pipe, *E*, enters the small airtight box or chamber, *T*. From the chamber rises a larger pipe, *F*, which, just above the top of the chamber, receives within it the smaller inlet pipe, *E*, and winds up through the protecting vessel concentric with the smaller pipe. On the second and third turns from the top the interior arrangement of the pipes is shown very clearly.

The operation is now clear. The air enters the pump at *G*, is forced through *H* and compressed, thereby being heated. The heat is removed in the cooling apparatus, *J*, and the compressed air, at the temperature of the water, goes on to *D*. There it descends in the inner pipe of the double coil, expands through *R* and is cooled thereby, passes through *T* and up through *F*, the outer pipe of the coil. There it cools the air in the inner pipe of the double coil. The air, therefore, reaches the valve, *R*, at a lower temperature than before, so that it is constantly falling in temperature, reaches *R* at lower and lower temperatures, and eventually the critical temperature of liquid air is reached and passed, and liquid air begins to collect in the chamber, *F*, as shown in the cut. By the faucet, *V*, it can be drawn therefrom as required.

If air is liquefied in the apparatus, every cubic inch of liquid represents about one-half a cubic foot of air withdrawn from circulation in the apparatus. Once the apparatus begins to liquefy air, it has to have new material supplied it, just as a grist mill needs a supply of grain to keep the stones in operation. A pipe at *A* connects with a second pump which pumps in new air as required, so as to maintain an advantageous pressure in the system—one which will give an economical relation between the pressures on the opposite sides of the aperture.

A minor yet important feature of this apparatus is that the liquid air collects at atmospheric temperature, or thereabout. The effect is twofold. It can be withdrawn much more easily than when it has to be taken from a receiver in which it is subjected to 50 or 100 atmospheres pressure. In the latter case it rushes out, only controllable by the faucet, and the mechanically atomizing effect plays a part in wasting it and facilitating its loss by gasification. But, stored under atmospheric pressure, it not only is quietly withdrawn, as required, but, by volatilization, it keeps its own temperature down. The maintaining it in a quiet state and in bulk operates to make it evaporate more slowly, the battle of the squares and the cubes, as it has aptly been termed, being involved.

It is evident that to make the difference of pressure $p^2—p^1$ (page 302) large, recourse may be had to the expedient adopted in steam engineering for expansion engines of high initial pressure. These are constructed with two cylinders (compound engines) or with three or more cylinders working in series, the steam passing seriatim from one cylinder into the next (triple, quadruple, etc., expansion engines). Just as in these engines the expansion is divided between several cylinders, so it is practicable in self-intensive refrigerating machines to force the air or gas through several apertures, letting each one take care of its fraction of the total difference of pressures, $p^2 - p^1$.

Linde has done this in a partial way in his laboratory apparatus, and the cut shows the modification

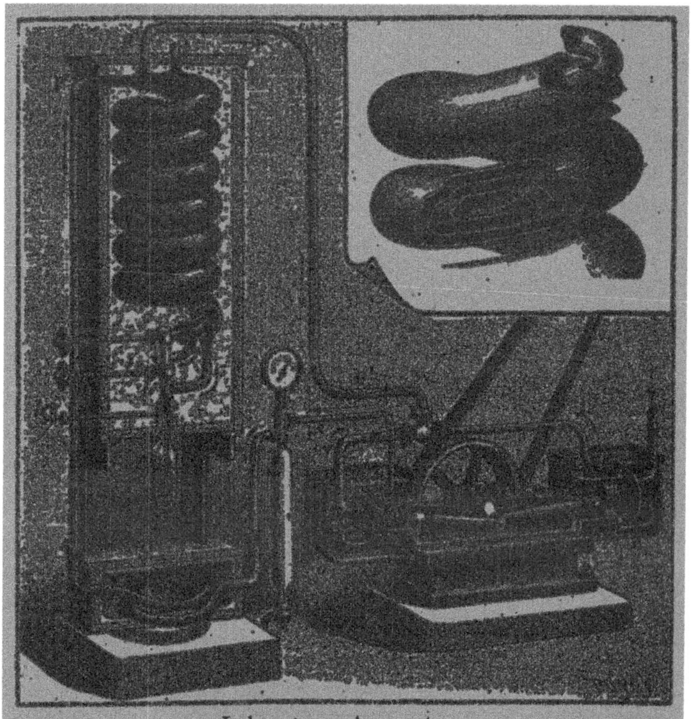

Laboratory Apparatus.

in question. If the description of the simple apparatus has been understood, the drawing alone will be almost self-explanatory. There are, however, various refinements introduced in this machine which need explanation.

A double-barreled pump is used which takes in air from the open room, the pipe on the right, with the arrow pointing down it, being the intake. The right hand pump cylinder pumps the air through the coil in the water jacket, *c*, and thence it passes into the cylinder on the other end of the pump. On its way to the other or left hand end of the double pump, it is joined by a stream of air from the interchanger or refrigerator, which air enters by the pipe, P^1. From the left hand pump barrel the air, now twice compressed, goes through a second water jacket, *d*, and by the pipe, P^2, passes to the left. These water jackets cool the air but partially. In order to more thoroughly cool it water is injected, and at *f* is a trap which removes most of the water. The air then goes through a coil in the small tank, *g*, which is surrounded by ice and salt. This cools the air thoroughly and removes the last of the water.

It will be remembered that in the first described apparatus an auxiliary pump was used to supply the deficiency of air, due to liquefaction of a portion thereof. In the laboratory apparatus the right hand pump barrel performs this function, compressing the air to 16 atmospheres only; the second or left hand pump barrel, taking in the air from the right hand barrel, and also the air from the pipe, P^1, compresses it all to 200 atmospheres.

The air thus compressed we have followed to its exit from the coil in *g*. Cold and dry, it rises to the top of the refrigerating case, entering it at P^2 and going down a spiral pipe. This spiral pipe is the inner one of a triple concentric coil, whose construction is shown in the small sectional cut in the upper right hand corner of the illustration. It descends through the interior coil to *a*, where it passes through an aperture regulated by a valve. Just below *a* is another valve, *b*. This valve is slightly opened, so that, of the air which passes *a*, one-fifth as near as may be passes *b*. The four-fifths of the air which does not pass through *b* rises through the annular space between the interior tube and the intermediate tube. This four-fifths of the air rises to the top of the refrigerating chamber and goes back to the pump by the pipe, $P^1 P^1$. This circuit is comparable to that in the first described machine.

The one-fifth of the air which passes through *b* has undergone a double expansion. It has expanded through two apertures, *a* and *b*. A portion of it when the liquefaction has begun passes on to the annular space between the intermediate pipe and the outside pipe of the coil, and, after passing through it, escapes into the open air at the top of the chamber. The outlet pipe is there shown leading from the outside tube up into the air. Three-quarters of it thus escape, one-quarter is liquefied and collects in the double-walled vessel, *c*. Thus, the air from the pump, entering the inner pipe at P^2, is cooled on its descent by the expanded air in the intermediate pipe. But this air is still further cooled by the constant uprising stream of still colder air rising in the outer pipe.

There is one peculiarity to be noted in the accumulative cooling action. The air from the pump entering at P^2 is working in the opposite direction to the colder air in the intermediate annular space or pipe. This is the correct method. But the cooling effect of the air in the outer tube is differently applied. This air rises, and cools in its rising the air in the intermediate tube, which is also rising. This is the wrong way of working, but its inefficiency is lessened by the fact that the entire quantity of air does not pass through the outer tube. It is only a question of one-fifth multiplied by three-quarters, which is three-twentieths of the original air. This is the quantity which passes up the outer tube. It operates, perhaps, more as a jacket than as a cooler.

The air, after it collects in the liquid state in the vessel, *c*, can be withdrawn by opening the cock, *h*. Enough back pressure is maintained in the vessel, *c*, to force the liquid air out at *h*, exactly like water from a soda water siphon.

It will be seen that the right hand pump barrel has to supply not only the deficiency in air caused by liquefaction of a portion of it, but has also to pump in air to supply the loss of that which escapes into the air after passing through the valve, *b*.

Another peculiar feature will be noticed. All of the air is not twice expanded. The majority is only once expanded, and all the liquid air which is produced is derived from the one-fifth of the total which is twice expanded through *a* and through *b*.

A pressure gauge is mounted on top of the trap, *f*, to enable the operator to maintain the proper pressure.

This apparatus, with the expenditure of three horse power, is credited with the production of nearly one quart of liquid air per hour.

The makers of liquid air, confronted with their great success, as yet scarcely know what to do with their wonderful product. One of their projects is to utilize it for the production of a highly oxygenated air, as it may be termed, for the production of a mixture of nitrogen and oxygen which will be very rich in oxygen.

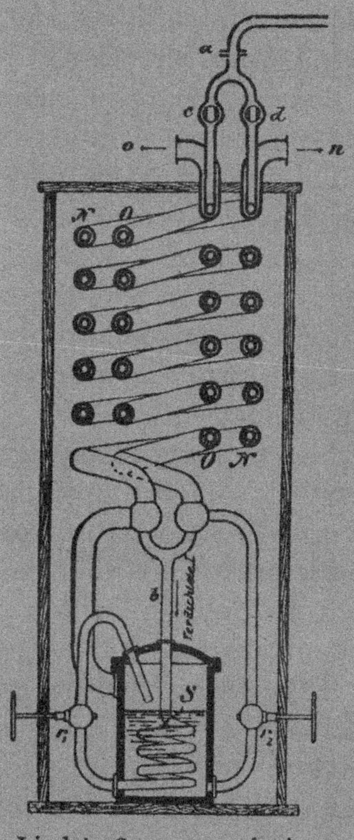

Linde's Oxygen-producing Apparatus.

The next illustration shows in diagram how Linde proposes to effect this by a continuous process. In the cut are shown a double set of annular or concentric pipes, forming two coils such as used in the first described apparatus. These coils are in parallel with each other. The air from the pump enters both coils by the small branched tube seen at the top of the apparatus and designated by *a*. It goes down the two interior tubes of the coils through the valves, *c* and *d*, and, leaving the outer concentric pipes, the tubes unite to a single pipe, *b*. Thence the single tube passes through the liquid air vessel, *S*, and emerges at the bottom. The air expands through the valve, r^1, and part of it liquefies and collects in *S*.

When air is liquefied and allowed to stand, it gives off nitrogen much more rapidly and in larger quantities than it does oxygen. Hence, a gas rich in nitrogen is given off by the liquid air in *S*, and this gas rises through the annular space between inner and outer pipe in the coil, which starts from the left of the liquid air vessel.

The liquid air, constantly growing richer in oxygen, passes out of a pipe leading to the right out of the bottom of the liquid air vessel and, controlled by the valve, r^2, evaporates into the annular space of the other coil. The nitrogenous gas in the one annular space and the gas rich in oxygen in the other annular space cool off the gas from the pump so as to form the true self-intensive heat interchanging system.

The two outer pipes are kept separate as they emerge from the interchanger. One, marked *n*, delivers a product poor in oxygen. This may be allowed to escape. The other, marked *o*, delivers a product rich in oxygen, which may be utilized for many technical purposes.

If the gases from the outer pipes of both coils are allowed to escape, one into the air, the other into an oxygen receiver, the pump will have to work upon new air constantly. There will no longer be a question of supplying a loss of a fraction of the air—all will have to be pumped in during the operation.

Linde's first successful experiments were performed in May, 1895. Fifteen hours' pumping was required to liquefy air, and then he collected some three quarts of liquid air per hour, containing about 70 per cent. of oxygen. He used in his interchanger iron tubes over 300 feet long, 1·2 and 2·4 inches in internal diameter respectively. His pump was a carbon dioxide or carbonic acid gas compressor, and he got from it a compression varying from 22 to 65 atmospheres. The liquid was crystal clear and bluish in color.

The inventor's own words describe his apparatus as eliminating heat from gas "exclusively by expenditure of internal work." This internal work he holds to be the work of separating the gas's own sluggish molecules from each others' vicinity.

CHAPTER XV.

THE HAMPSON APPARATUS.

Hampson's apparatus—Its general features of construction—The jet and regulating device—Thermal and mechanical advantages—Data of its operation—Use of cylinders of compressed gas instead of pumps—Application of preliminary cooling to the air or gas to be liquefied.

The Hampson apparatus is the invention of Dr. W. Hampson. It is very simple and resembles very much the Linde apparatus, and it works precisely on the same lines.

The cut shows a section of the apparatus. A cylindrical case is lined with non-conducting material. It contains three coils of pipes, each coil consisting of a single range of pipes arranged almost in the shape of a cylinder. The coils of pipe are laid in what may be termed the grooves of helices or screws, formed by winding partitions whose course is parallel with the axes of the coils of pipe, so that the section of the apparatus shows the circular tube sections, each in a little square. The perspective view of the end of the innermost coil on page 322 shows how the pipes and partitions are disposed.

The air enters by the small tube at the upper right hand portion of the case. It goes down the long outer helix, passes to the bottom of the intermediate one, and rises through its coils to the top. Here it passes into the central coil and descends to the bottom of it, near the lower end of the liquid air reservoir. The air here issues through a jet into the body of the apparatus. It follows the course of the helically bent pipes; first up the center, then down the intermediate chamber, and then up the exterior chamber, escaping at the larger pipe. Its course, it will be observed, is exactly contrary to that followed by the air on its journey within the pipe. The helical partitions guide it on its return course.

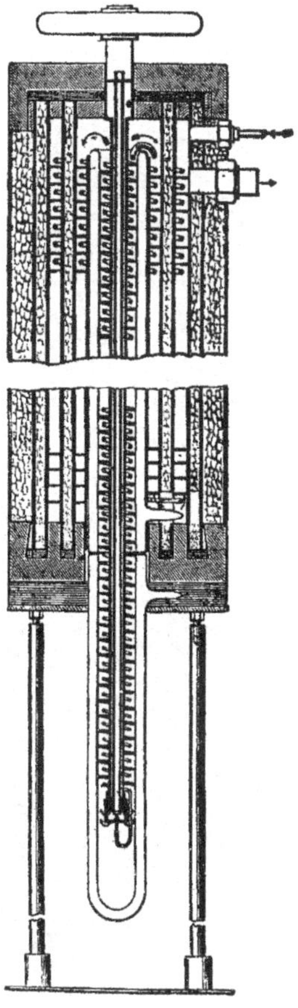

Hampson's Gas Liquefaction Apparatus.

The jet through which the gas expands is shown in the next cut. Its delivery capacity is regulated by screwing toward its face or away from it the flat, or nearly flat, piece shown. The smaller its delivery capacity at a given pressure, the greater is the difference of pressure or degree of expansion which it establishes at any pressure.

In illustration on this page, showing the internal arrangement of the coils, it will be seen that the upturned jet points to the center of a threaded aperture, a pipe from which extends to the top of the apparatus. Through this aperture a long stem passes, with a screw near its bottom and an almost flat end. By screwing the rod up or down, the flat end is brought nearer to or withdrawn from the jet, as described, the delivery of the aperture is made greater or less, the whole operating as a regulating valve, and, there being no interior parts, the chance of any obstruction is minimized. The valve rod is shown in place in the cut showing the full apparatus.

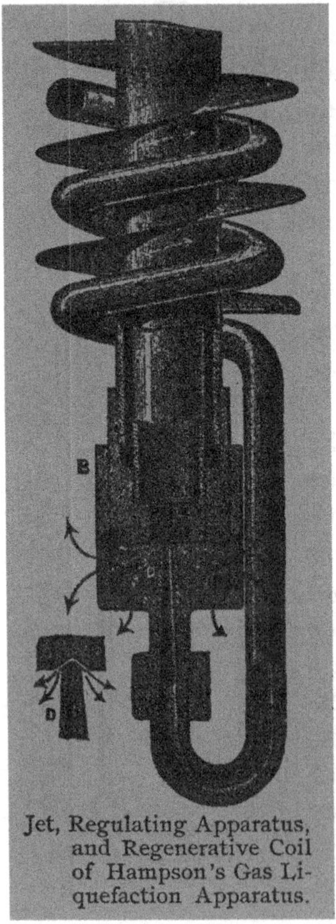

Jet, Regulating Apparatus, and Regenerative Coil of Hampson's Gas Liquefaction Apparatus.

The pipes are made as thin as possible, in order to facilitate rapid and efficient cooling. The compressed and the expanded air are in finely subdivided volumes, so that they readily interchange temperatures, and the long and devious course in opposite directions, followed by the two divisions of the air, conduces to the same end.

The action has been fully explained already in the description of the Linde machine. The compressed air expanding becomes cool. The cooled gas following the coils cools the air within them. The temperature constantly falls, and presently liquefaction occurs. The liquid air collects in the reservoir below the main case.

The apparatus is operated by a compressor or by the use of cylinders of compressed air.

The compressor must deliver air at a pressure of 80 atmospheres or over. An engine power of 3·5 horse power is required to drive the compressor, and about 1·2 quarts of liquid air are produced in an hour. No preliminary cooling is required.

If the compressor delivers air at a pressure of 120 atmospheres, air begins to liquefy in 16 minutes; if at a pressure of 130 atmospheres, only 10 minutes are required.

If a cylinder of compressed oxygen is used instead of the pump, the conditions are less favorable, as the pressure constantly falls. Cylinders adapted for the purpose can be procured. When such are employed, an auxiliary cylinder of liquid carbon dioxide is needed. This is used to cool the apparatus preliminary to the admission of oxygen. The latter is compressed to 120 atmospheres. One hundred and twenty-five cubic centimeters can be collected therefrom.

The preliminary cooling by the carbon dioxide is effected by passing the gas, intensely cold from its gasification, in at the bottom of the apparatus, so that it follows the general path followed by the escaping air or oxygen in the regular operation of the apparatus.

It will be seen that the idea of circulating the identical air over and over again is not carried out. All that does not liquefy escapes. But this is merely a detail. If oxygen or any expensive gas were being condensed, the cheapest way would be to use it over and over again, and this could readily be done by a compressor with its inlet connected to the outlet of the apparatus.

There is one important point to be considered in working with a compressor as contrasted with the use of a cylinder of compressed gas or air. The action of the compressor heats the gas or air; so it is advantageous to cool it by water, or otherwise, before admitting it.

But if a cylinder of compressed gas is used, there is no heating. There is even a reduction of temperature, due to expansion; so that an advantage is gained.

This applies to any similar liquefaction apparatus.

In Dr. Hampson's laboratory apparatus the liquid air or oxygen can be withdrawn from its recipient by siphon, or the receiver can be removed with its contents by unscrewing a vulcanite cap at the bottom of the apparatus.

The disposition of pipes varies somewhat in different types of apparatus, but the same principle is followed in all of them. The great object to be attained is lightness of the interchanging system of pipes, in order to increase thermal conductivity.

CHAPTER XVI.

EXPERIMENTS WITH LIQUID AIR.

Experiments with liquid air—Formation of frost on bulbs—Filtering liquid air—Dewar's bulbs—Liquid air in water—Tin made brittle as glass—India rubber made brittle—Descending cloud of vapor—A tumbler made of frozen whisky—Alcohol icicle—Mercury frozen—Frozen mercury hammer—Liquid air as ammunition—Liquid air as basis of an explosive—Burning electric light carbon in liquid air—Burning steel pen in liquid air—Carbon dioxide solidified—Atmospheric air liquefied—Magnetism of oxygen.

We shall now describe some of the lecture experiments with liquid air. These are generally reproductions of experiments shown by Charles E. Tripler at his lectures and demonstrations. For most of the illustrations our thanks are due to the *Scientific American* and to *McClure's Magazine*.

When liquid air is poured into a glass flask it boils energetically, and the outside soon becomes covered with hoar frost, and clouds of moisture condensed from the atmosphere descend from it. From its mouth the same cloud is seen apparently escaping. But this cloud has nothing to do with the liquid air itself. It is simply the moisture of the atmosphere condensed by the cold of the air as the latter evaporates from the liquid state.

By courtesy of *McClure's Magazine*. Copyright, 1898, by The S. S. McClure Company.

Filtering Liquid Air—Frost-coated Bulb.

The above cut shows the filtration of liquid air into a Dewar bulb. Ordinary filtering paper is employed, and the solid or cloudy matter, such as solid carbon dioxide, is effectually removed, and a beauti fully clear bluish liquid drops into the bulb. The bulb on the right is one just showing a coating of hoar frost.

If a Dewar bulb is substituted for the flask, the air lies comparatively quiet. In a good bulb only one or two tiny threads of bubbles rise through the liquid, reminding the observer of champagne whose effervescence has nearly exhausted itself. On first introduction the liquid air may be quite agitated and steam may appear escaping from the neck.

On dropping liquid air into a flask of water, the action is very violent. The air at first is lighter than water, but it grows heavier as it loses nitrogen. It sinks, after a little, partly gasifies, floats up, and forms ice about itself, and at last disappears. A larger vessel of water than is indicated in the cut may be advantageously used. The small cut gives an almost conventional representation of what occurs when liquid air is poured into a narrow-necked flask of water. In the actual experiment, which is best performed in a wide-mouthed bottle of water, there is much agitation and disturbance. The globules rush about, vapor forms about the mouth of the vessel, and

the appearance which is so well presented in the cut below is seen.

Many substances are made brittle by immersion

By courtesy of *McClure's Magazine*. Copyright, 1898, by The S. S. McClure Company.

Liquid Air in Water.

in liquid air. We have seen that lead becomes elastic and that the pitch of a tuning fork is raised by immersion. It is quite possible to make a tuning fork out of soft metal which will become resonant on immersion for a few seconds in liquid air. A tin dipper after a few minutes' immersion becomes almost as brittle as glass and is broken by a blow.

India rubber, such as children's balls are made

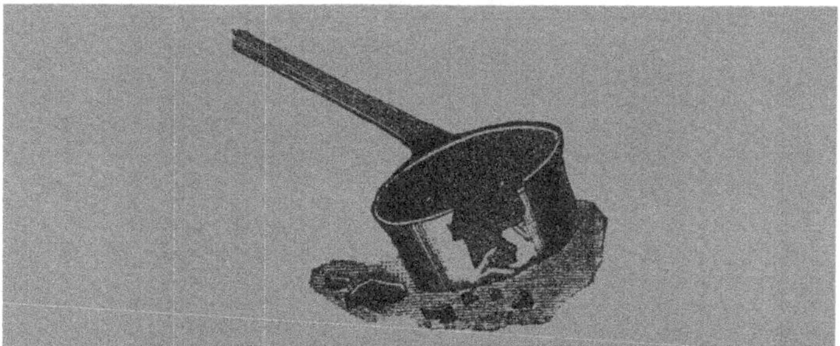

of, becomes almost as brittle as glass after floating a few minutes in it. The cut showing a ball in liquid air brings out another point of interest—the formation of the cloud of moisture and its descent.

The air which volatilizes from the liquid air is very cold and pours over the top of the vessel like water and carries the cloud with it. The cloud is composed of moisture condensed from the outer atmosphere.

The freezing of an alcoholic liquid gives a good proof of the low temperature of liquid air. Liquid air is poured into a glass of whisky or alcohol, and the liquor freezes. The cut shows a sort of icicle

By courtesy of *McClure's Magazine*. Copyright, 1898, by The S. S. McClure Company

Alcohol Icicle.

of alcohol lifted up on the end of a rod out of a glass of alcohol thus frozen.

A test tube containing liquid air is placed in a glass of whisky. The latter is soon frozen solid, and can be lifted out of the tumbler in a lump. On standing a few minutes after the air has evaporated, the test tube can be taken out, and a sort of tumbler whose material is frozen whisky is produced.

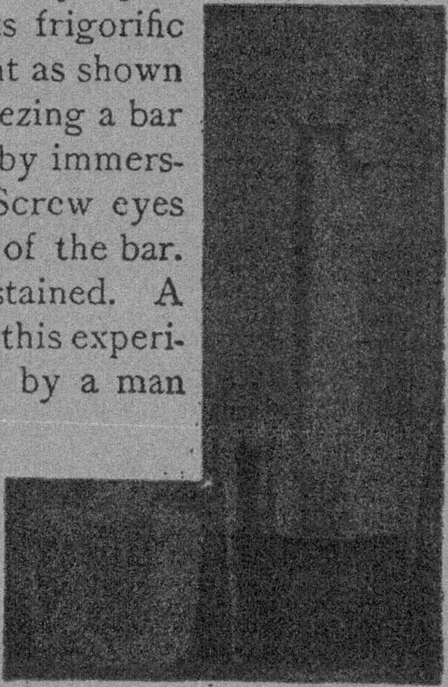

Mercury is often frozen by liquid air as an example of its frigorific power. The experiment as shown in the cut consists in freezing a bar of mercury in a mould by immersing it in liquid air. Screw eyes are frozen into the ends of the bar. A heavy weight is sustained. A striking presentation of this experiment has been effected by a man

hanging from such a bar of mercury. Another example of the effect of cold upon mercury consists in making a tuning fork out of it. It is easy to see that the changes which may be rung upon this phase of low temperature are very numerous.

Another experiment consists in casting a hammer head out of mercury. A mould is prepared with a handle thrust into it, and mercury is poured in. Liquid air is poured upon the mercury. After a few minutes' standing the mercury freezes so hard that it can be withdrawn from the mould, and a nail can be

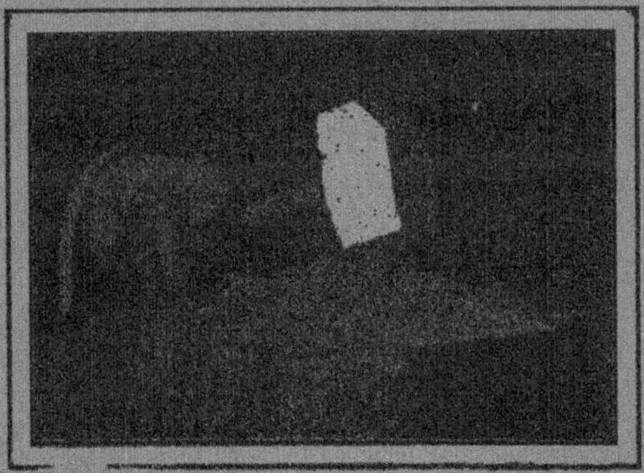

By courtesy of McClure's Magazine. Copyright, 1898, The S. S. McClure Company.

driven with it. We are not aware that a mercury nail has ever been driven into wood.

The gasification of liquid air is nearly irresistible in the pressure it produces when confined. A quantity is poured into a metal cylinder closed at the bottom and a plug of wood is driven into the top. In a few seconds the plug is expelled as if by the explosion of gunpowder, with a loud report.

If a piece of paper is saturated with liquid air and lighted, it burns with much energy. The longer the liquid air has been kept, the more violent is the combustion. The standing of the air causes it to grow richer in oxygen. A piece of boiler felt, which

ordinarily cannot be made to burn, if saturated with liquid air rich in oxygen, burns most brilliantly, and

if liquid oxygen is used, almost explodes. This is in the air. If confined, a violent explosion ensues.

An electric light carbon brought to a red heat and plunged into the liquid burns beneath it. The carbon dioxide formed by the combustion remains in great part in the liquid, freezes solid and sinks to the bottom.

A steel pen or a watch spring can be burned in liquid air which has been kept standing a few minutes. A bit of sulphur may be placed on the end of the steel and ignited to start the combustion. An interesting variation on this experiment is to place the liquid air in a tumbler made of frozen whisky, as described on page 330. The pen or watch spring is burned in this. The white heat of the burning pen, the intense cold of the air, and the alcoholic liquid hard frozen form a set of incompatibles which it would be hard to equal. The combustion of steel, a metal once supposed to be incombustible, is occurring more vividly than that of the most familiar inflammable substances and in a vessel made of a frozen liquid once supposed to be incapable of congealment. The material of the pen is practically that out of which grates and stoves are made. The material of the tumbler is approximately one-half alcohol, which latter liquid has long been used to prevent freezing.

A kettle of liquid air placed on a cake of ice boils actively because of the heat of the ice which supports

By courtesy of *McClure's Magazine*. Copyright, 1898, by The S. S. McClure Company.

it. If the boiling is not rapid enough, it may be accelerated by adding ice water or even a lump of ice to the kettle. This shows that ice is hot.

If carbon dioxide gas is directed by a jet upon liquid air, it is liquefied and also forms carbon dioxide snow.

But far more impressive than this is the experiment illustrated in the diagram, which is self-explanatory. A tube of liquid air is connected to an air pump and exhausted. The cold is so intense that, after a few minutes, liquid air drips off the outside of

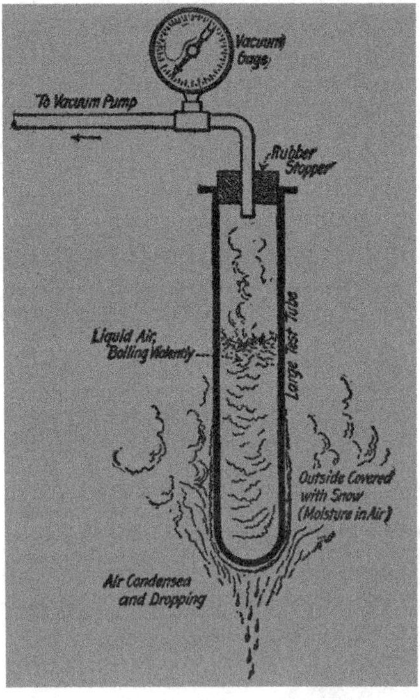

the tube. This is the air of the atmosphere reduced to the liquid state by the intense cold of the tube, due to the boiling of the air within it.

The phenomenon reminds us of Dewar's experiment with liquid hydrogen, whose cold was so intense that it liquefied the atmospheric air. It is also useful in bringing before us the dependence of liquefaction upon temperature and its independence of pressure.

Oxygen was discovered to be diamagnetic by Faraday. A tube with outlet is filled with liquid air and is suspended by a thread as shown. A powerful magnet attracts it as if it were a bar of iron or steel.

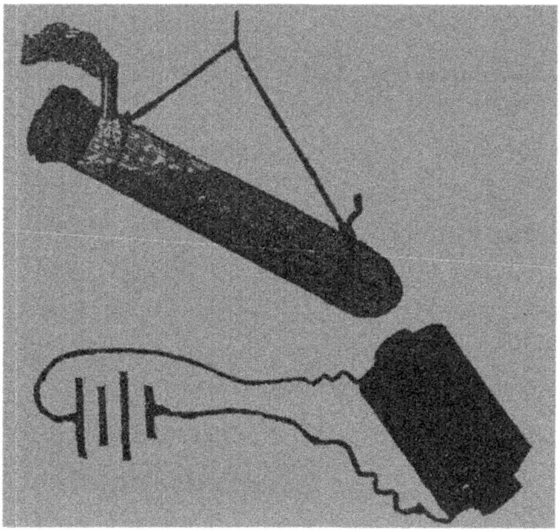

This is an incomplete presentation of the experimental side of our subject. Changes in colors of chemicals and many other phenomena can be shown. The description falls far short of the actual witnessing of the experiments.

CHAPTER XVII.

SOME OF THE APPLICATIONS OF LOW TEMPERATURES.

Frigotherapy—The frigorific well—Pictet's experiment—Effects of the first trial of the system—Medical uses of liquid air—Critical point as test of purity of chemicals—Purification of chemicals by low temperature crystallization—Low temperature distillation—Regulation of chemical reactions by cold—Liquid air explosives—The principle of their action—Liquid air in electric power transmission—Liquid air as a reservoir of energy.

Prof. Raoul Pictet has during the last few years given much attention to the uses of the intense cold produced by the application of liquefied gases. The purification of chemicals, the testing of the same for minute quantities of impurities by intense cold and by the observation of the critical point, and the regulation of reactions, are included in the scope of his work. Another of the uses to which he proposed to put the application of intense cold is the treatment of disease.

He conceived the idea that simple exposure of the system to a very low temperature for a short time might be productive of important effects. The human system in the Arctic regions has endured very low temperatures without any effect upon the personal hygiene as far as discernible; but it remained to be seen whether, by descending far below these natural extremes, a constitutional effect could not be produced.

He constructed what he termed a frigorific well, a small chamber, double walled, and lined with thick non-conducting material, to protect the subject from contact with the walls or floor. Such well was about 6 feet deep and 2 feet in diameter. By use of the cold derived from the *liquide Pictet* (page 169) the temperature within the well could be reduced to —110° C. (—166° F.) A foot stool was placed upon the floor. This was so arranged that the patient could stand upon it, with his head in the open air. A woolen cover was thrown over his shoulders, so that the head alone emerged, and the rest of the person was immersed in the chilling atmosphere as if in a cold bath. The clothing was not removed. The chill penetrated it readily.

The effects of the immersion were very marked. The body had to maintain its heat, and this can only be done by a more vigorous process of oxidation. As Prof. Pictet expresses it, the body becomes autophage or self-devouring. The temperature taken by a thermometer in the mouth rises in amount from 0·2° to 0·9° C. (0·36° to 1·6° F.) The temperatures of the human body, it will be remembered, are always expressed in this country in Fahrenheit degrees, so that the above temperatures are expressible as 98·76°–100° F., taking 98·4° F.° as the average human temperature.

A slight feeling of epigastric constriction is sometimes felt by the subject, a slight momentary paralysis in the lower extremities may be experienced, but all is quickly succeeded by a feeling of general invigoration. A reaction generally occurs before the patient leaves the well.

After a while the temperature falls below the normal, and a slight vertigo may appear and the pulse may slacken.

A two hours' exposure proved fatal to a dog.

Pictet himself reports that in his own case he effected a remarkable cure by the use of the cold well. He had suffered for years with stomach trouble of the dyspeptic type, and resolved to try the effect of extreme cold upon himself. His respirations were at the rate of fifteen and one-half per minute; his pulse beat at a frequency of sixty-three.

He descended into the cold well, wearing a heavy wrap. A plank lay upon the bottom for him to stand upon. In order to keep in motion, he lifted his feet successively six inches high, with a frequency of forty-two per minute. For four minutes no especial sensation was experienced. After five minutes, or thereabout, an indefinable sensation was felt, and a desire for nourishment appeared, marking the beginning of what he terms a *frigale.* The pulse beats rose in frequency to sixty-seven per minute, and the respiration to nineteen. Each respiration was deeper than usual.

After eight minutes' exposure he emerged, feeling a sort of prickling sensation all over the body, but no cold affected the skin. A well defined hunger was present, almost disagreeable in its craving effect.

On walking homeward, after two or three minutes a reaction set in, exceeding in intensity that due to a cold bath. The body seemed penetrated by a myriad of fine needles. He states that this expression gives but a feeble idea of the physiological consequence of the restoration of the normal circulation. The reaction lasted at least fifteen minutes.

This was on February 23, 1894. He states that on that day, for the first time in six years, he ate a full meal with enjoyment.

During February and March of that year he made eight experiments in the descent into the cold well. The periods varied from eight to eleven minutes each. The same sensations and reactions accompanied each trial. He gained weight rapidly after the treatment, and found his health radically improved.

In the year 1895, at Geneva, Pictet was invited to exhibit his work before the National Exposition. Among other things, he installed two cold wells which could be brought to a temperature of —110° C. (—166° F.)

The apparatus was placed in charge of two physicians, Drs. Cordes and Chossat.

The wells were thoroughly protected by fur. They were entered by a ladder or the patient was lowered into them by ropes. Footstools of various height were provided, so that patients, whether tall or short, could be properly immersed. A woolen covering was provided for the shoulders.

The working temperature rarely rose above —90° C. (—130° F.), and was often much lower.

It became quite the fashion to take a cold air bath. So many presented themselves that the physiological examinations were somewhat restricted. The desire on the part of the management, however, was to facilitate the trial of the cold air wells by as many patients as possible.

The patients were examined carefully in many cases; the temperatures were taken before and after an exposure of ten or twelve minutes. In a few instances the exposure exceeded fifteen minutes. Some visitors descended only once, others a dozen times.

Full reports on the subject will be found in *Science Français* of November 6, 1896, and a report was presented to the Medical Academy of Paris by Dr. Cordes at its meeting on October 29, 1897. Finally, a most elegant presentation of the subject is given in Prof. Pictet's book "La Frigothérapie, ses Origines, son But," Paris, 1898. Curves indicating the changes of pulse frequency and of temperature, with other observations for ninety-seven cases, are given.

A method of quickly applying frigotherapic treatment locally is due to Dr. Ribard. He uses solid carbon dioxide alone or mixed with ethyl chloride as the source of cold. This he applies locally to the skin, protected by felt.

Dr. G. Fish Clark, of New York, writes that he has removed cancer, certain forms of bunions, corns, warts and superfluous hair by means of this agent. The tissue, when the air has thoroughly worked upon it, is practically cut off by means of a temporary status in the circulation of the blood. The circulation is not renewed if a certain amount of care, obtained by experience, is taken, as may be indicated in each individual case. The parts beneath the morbid tissue or morbid growth not affected by the low temperature of the liquid air are held intact, and use their circulatory system by means of anastomosis and returning of arterial blood (after it has become deoxidized) to the veins by means of infiltration through the interstitial spaces. This process forms a new skin surface under the morbid and frozen surface.

The result is an upheaval of the super-tissue, which, as it dries and shrinks, eventually falls off like a scab. The process of applying must be studied, and it is dangerous to place it in inexperienced hands, as the freezing of vital organs, the danger of involving large distributing arteries and veins, and the involvement of osseous tissue, must be avoided. It must be determined accurately by the physician how deep an application is going in a certain interval of time.

From his own observation, he has failed to draw the same conclusion as to its effect upon bacteria as M. D'Arsonval, Paris, has arrived at. He has, as far as he has investigated, found an utter destruction of microscopic life. He has not, however, experimented with the bacilli D'Arsonval used. He affirms that he has the greatest faith in liquid air as a means whereby humanity will receive great aid, and that in many cases where the knife is now used this agent will be found a most welcome substitute. The pain in its application is at no time sufficient to require an anæsthesia, it is complicated with no hemorrhage, and the patient, after its proper application and dressing, feels no additional inconvenience. "If by inventing this process of manufacturing liquid air Mr. Tripler has accomplished nothing else than this, his name will be treasured at least in medical history as that of one of its most valued contributors."

Pictet has applied a curious observation which he

made in determining whether chemicals are pure. He found that an infinitesimal amount of impurity, while it affected the boiling point very little, would make a difference from ten to sixty times as great in the temperature of the critical state.

An apparatus was made by which a group of tubes of various liquids could be heated to known temperatures under observation. The disappearance of the meniscus was taken as indicating the critical state, which was supplemented by the nebulous effects which occur at the same point.

To chloroform were added a few drops of alcohol. The boiling point was barely affected, but the critical temperature was changed several degrees. A number of other chemicals were tried with analogous results. For a certain class of substances, therefore, a delicate test of purity exists in the determination of the critical point.

The great degree of cold which the liquefaction of gases puts in the chemist's hands extends an old time method of purification to new fields. For generations past crystallization has been the great agent in purifying salts. If a chemical salt is dissolved in water and the solution is evaporated down until it becomes a relatively strong one, a point is often reached at which the dissolved substance tends to separate. With the majority of salts this point is attainable. If the strong solution is left to stand, the salt will gradually separate in crystalline form.

The phenomena above described in a few words naturally do not occur with all substances, not even with all soluble substances. Again, the phenomena are not restricted to solutions in water; they may occur with other solvents. But water is the great solvent, and we are more familiar with crystallizations from water than from other substances.

In all nature there is no more wonderful example of mathematical exactitude than that supplied by the laws of crystallization. The forms of the crystals are based on exact laws formulated originally by the Abbé Hauy.

The fact that a crystal is an exact, mathematically determined form almost implies that when a substance forms a crystal it must be a pure substance. If the substance dissolved were dirty and impure, crystallization, should it occur, would have at least a tendency to purify it.

An impure substance is dissolved in water, is crystallized therefrom, the crystals are removed and drained from the liquid—mother liquor, it is called—and are found to be much purer than was the original substance. They may be redissolved and recrystallized, when the second crystallization will impart a still higher degree of purification.

This process has long been employed by chemists and manufacturers to purify salts, and is still the great process used to obtain pure chemicals.

When water is exposed to cold it solidifies, and its solidification is a species of crystallization, although the crystalline formation is, as a rule, not visible. It is brought to view by melting ice under proper conditions, and all are familiar with the beautiful crystalline forms which are discernible in snowflakes. We should expect, therefore, that the freezing of water would have a purifying effect upon it.

It has this effect. Ice is purer than the water from

which it is made. If cider is exposed to cold, the water freezes out in a relatively pure condition, and the cider is left as a sort of mother liquor, so much stronger than before that what is almost a brandy results. Here the cider constituents are the impurities, and they are left in the mother liquor in greatly concentrated condition, and the water is crystallized out as ice in a relatively pure state.

Water is ordinarily purified by distillation. It would be perfectly practicable to purify it by repeated freezings, if distillation could not be effected.

Liquefied gases, by their innate cold and power of absorbing heat energy or rendering heat latent, extend the range of the freezing processes to new fields. Liquid air can solidify and thereby purify alcohol.

A number of very important chemicals can be purified by intense cold. One of the most familiar of these is chloroform. Used as an anæsthetic, it is uncertain how much of the bad effects of chloroform are due to its impurities. Irrespective of any danger to life, there are after effects which it is desirable to overcome or minimize. The purer it is, the less are these after effects, and it is quite possible that with absolutely pure chloroform, deaths of patients from the effects of its administration would be far less frequent than they now are.

Chloroform is purified by freezing. On subjection to a proper degree of refrigeration, the pure chloroform crystallizes out almost like sodium sulphate from water. The very cold crystals are removed and melt, and an extremely pure product is the result. The process is termed one of rectification at low temperature, and can be applied to a number of liquids. Chloroform is taken as a typical substance, and as one for which a great demand exists. Ether is another chemical product which is thus purified with success, and alcohol can be purified by the freezing process until it is 100 per cent. pure, or is what is known as absolute alcohol. Various anæsthetics are purified by freezing.

Formerly these methods were inapplicable, simply because the degree of cold requisite for their execution was unattainable.

Distillation by heat is attended with the objection that the heating may impair the product. Low temperature distillation is made practicable by utilizing the intense cold of liquefied gases to condense the distillate. In this way so high a vacuum is produced that a liquid will distill with relative rapidity at ordinary temperatures. It is a reversal of the ordinary course of operations. Instead of applying heat to the retort and forcing off the gasified liquid against the pressure of the atmosphere, the latter is removed and the gases which take its place are condensed by intense cold, so as to maintain an almost perfect vacuum over the liquid, which distills without artificial heat.

Chemical reactions are so greatly modified by temperature that the cold of boiling liquefied gases may bring about radically different results in these cases. Thus, if organic substances are treated with nitric acid, the products will vary according to the temperature at which the interacting substances are kept. As illustrations of compounds produced by the action of nitric acid on organic substances, nitroglycerine, guncotton and many similar substances may be cited. These have extensive uses as explo sives, and by deoxidation give a host of products such as the aniline dyes. Any process which affects these reactions would affect the most important field of chemical industry. Heat, in the popular sense, has hitherto been the great agent in producing chemical reactions and in modifying them. Intense cold may now be looked on as a supplementary agent.

An explosive is a substance whose action may depend upon various chemical and physical actions. If two volumes of hydrogen are mixed with one volume of oxygen, a colorless mixture of gases results. If an electric spark or other source of heat is applied to the mixture, they at once combine suddenly, and with production of great heat. The result is an explosion, and the operation of combination produces a sound like a pistol shot. The mixture can be made to discharge a shot from a gun or to blast rocks.

Another class of explosives operate by simple breaking up of a feeble chemical combination. Chlorine and nitrogen can be made to unite and produce an oily liquid—a chemical combination of one atom of nitrogen and three of chlorine. On the least disturbance, or without any apparent reason, the compound will explode, simply reproducing chlorine and nitrogen. But, simple as it seems, the explosion is of fearful violence, and it is truly appalling to read of Davy's and Faraday's work with this substance, one of the most dangerous known to humanity.

It is unnecessary to go further. When a substance can be made in which a very violent chemical action can be induced, the heat produced and the changes in volume may be so sudden and great that an explosion results. Such a substance is termed an explosive, and there are a great many of such now in service.

One of the proposed uses of liquefied air is as a constituent of an explosive. If air is liquefied, it occupies about one eight-hundredth of its former volume, so that there is involved in its liquefaction a concentration of its oxygen to that extent. Then, we know that, by standing, the nitrogen evaporates more rapidly than the oxygen, so that a constant action of enrichment in oxygen is taking place as regards the unevaporated liquid. Thus, the liquefaction of air and subsequent enrichment may amount to a concentration of its oxygen of sixteen hundred or more times.

Even this is not so remarkable as it might seem. We are very familiar with oxygen in liquid and solid form in combinations of the chemical order. Thus, water, which we know most familiarly as a liquid, or as a solid, contains eight-ninths its weight of oxygen. Startling as it seems, it is no paradox to say that water is approximately pure liquid oxygen. This assertion would be based on its chemical composition by percentages or proportions by weight. But there is more than this to be looked at. By its affinity for hydrogen it is locked fast in the water molecule, so as to be comparatively inert. Those who have seen the fierce combustion produced by soaking organic matter in liquid air and then igniting it would never think of employing it as a material to put out fires. Yet we use water for this purpose, although it is far richer in oxygen than is liquid air.

Under certain conditions water can support combustion. If steam is passed through a mass of red hot copper borings, iron borings, coal and many other substances, it gives up its oxygen to them, the hydrogen severs its alliance, and a true combustion ensues at the expense of the oxygen of the water.

It is hard to bring about a combustion in water vapor, and in liquid water it is all but impossible, owing to its cooling powers.

The air we breathe contains about one-fifth of its volume of oxygen, and fires burn in it with far greater energy than in steam, which contains one-third its volume of the same gas. This is because oxygen in air is free and uncombined, and can unite with anything that claims it, without having to dissolve any bonds which unite it to other elements.

We are familiar with oxygen in the solid state in innumerable compounds. For purposes of combustion and explosion, we select those that are richest in oxygen and which have it most feebly united or combined. The "villainous saltpeter," potassium nitrate, contains in round numbers 48 per cent. by weight of oxygen, which is very feebly combined, and is, therefore, so ready to combine with carbon, sulphur and other compounds that for centuries it has figured as an ingredient in the great explosive gunpowder, which has ended many a life on the battlefield, a service some may be weak enough to consider of very questionable utility. The scientist cannot but consider the human body as a very exquisite mechanism, and must regard its destruction by one who cannot adjust and create its mechanism as a work opposed to every ethic of true science. Science always contains for its true votaries elements of admiration and wonder. Destruction of that which cannot be created or resynthesized is an abject confession of weakness that should be most discordant with every note of the scientific student's nature.

Now take liquid air which by standing has become rich in oxygen. It is liquid and of about one-third the specific gravity of typical solid oxygen-containing compounds. One-half of its weight may be oxygen which is absolutely free and uncombined, ready on provocation to unite with many elements without having any bonds of union to sever. It is evidently an available substance for a constituent of an explosive or for an inciter of violent combustion.

It is found that if liquid air, after standing a little while, so as to evolve nitrogen and become rich in oxygen, is poured upon organic matter, such as cotton, felt, powdered charcoal and similar substances, a violently combustible product is formed. A piece of heavy felt which can hardly be induced to burn in the open air, when soaked with liquid air, burns with the brilliancy of a piece of pyrotechnics.

This is combustion. Rapid combustion is explosion, and with such mixtures explosion can be brought about by confinement before ignition and by ignition with a detonator. The shock and heat set the whole off at once, and an explosion comparable to that of gunpowder results.

The following are the general features of Dr. Linde's practical trials of the liquid air explosive for blasting rock and coal: Charcoal is broken up into grains about the coarseness of beach sand. The effect of pouring liquid air upon the porous mass with its many points is to eliminate the spheroidal state and to provoke violent ebullition. This would be so great as to scatter the charcoal to right and left. Accordingly, to keep it together, the charcoal is mixed into a sort of sponge, with one-third of its weight of cotton (cotton wool or waste).

Liquid air, which has stood long enough to contain about half its weight of oxygen, is poured upon the mixture of wool and charcoal. An ebullition at first occurs, during which more nitrogen than oxygen goes off, and a further concentration of oxygen is effected. The moist mixture is rapidly charged into insulated paper cartridges, and is ready for use within five or ten minutes. It must be at once placed in the shot holes and exploded by a detonator, preferably an electric one. But any detonator which can be rapidly exploded will answer. Delay is fatal in one sense—it destroys the efficiency of the cartridge. After fifteen minutes to half an hour the liquid air will have so completely evaporated that no explosion can be produced.

This might seem a defect, but it is quoted as a merit. Countless accidents have happened in mining and tunneling operations from cartridges hanging fire, as it is called, in blast holes, only to go off unexpectedly, and killing and maiming the workmen. Half an hour after a liquid air cartridge has been placed in the hole it is innocuous.

By using air which has stood a longer or shorter time, the power of the explosive and the heat produced in its explosion can be controlled at least to some extent, even if it must be considered largely guesswork.

The explosive was used for several months in a coal mine at Pensburg, in Bavaria, near Munich, with good results. Where power costs nothing the explosive is a very cheap one. In tunneling operations it often happens that there is a surplus of power derivable from streams that flow in the vicinity. The European engineers show a great aptitude for utilizing such sources of energy. Where such are available, this would be the cheapest possible explosive, as well as the safest.

In America, Tripler has experimented in this direction, and has found that he could blow heavy steel tubes open as if with dynamite.

Elihu Thomson presents the possibilities of liquid air in electric power work. Few realize how large an item capitalization plays in the problem. The installation of a long line of copper is an expensive matter, and successful efforts are made to reduce it by employing high potential difference. But could the temperature be reduced to that of liquid air, a thin wire would carry a large current at relatively low potential difference, or at the high potential difference a very much larger one. As far as the cost of copper went, the capitalization of the line would be slight, in proportion to the power transferred. There would be every excuse for an expensive construction of a line which would carry a large current. The capitalization per unit would be quite small.

The idea of Elihu Thomson is expressed by reference to the power of Niagara Falls. An expensive power installation is there established which works to its full capacity for only a little over one-third of each day. He suggests that the power might be used during the night hours for making liquid air which could be stored in tanks well insulated from the outer air temperature. The inevitable evaporation of air could be utilized to perfect the heat insulation by being led down through the jacketing of the tank.

A furnace in a steel works or other industrial establishment may have a temperature on its hearth and working chamber of two or three thousand degrees above that of the air, yet there is no difficulty in insulating it by a firebrick lining and, perhaps, ordinary brick exterior, so that the hand can be placed upon the outer surface without being burned. Between liquid air and the atmosphere there is but one-eighth the difference of temperature that exists between the heat of a furnace and that of the air.

The copper conductor could be inclosed in a pipe which could be kept cold with liquid air. Such a line need not involve a loss in the energy transported of more than one or two per cent. In most long distance lines a loss of ten or fifteen per cent. of the energy is allowed for. It is possible that the saving of most of this might pay for the cost of liquid air, irrespective of the increased capacity of the line.

A few years ago it would have seemed absurd to make such a suggestion. But there is not a particle of absurdity in it. The achievements in the production of liquid air by Tripler and others, and the carrying of it hundreds of miles by rail in jacketed buckets, show how easy a substance it is to handle, once a sufficient quantity is brought together.

The surfaces of solids of identical shape vary with the squares of their linear dimensions. Thus, if there are two of Tripler's air buckets exactly alike, except in size, and if one is twice as large as the other, the surface of the tin and of the open top will be four times as large in one as in the other. The volume varies as the cube of linear dimensions. Therefore, in the case cited, the larger bucket will hold eight times as much liquid air as will the smaller one. Therefore, if we state the relation of surface to volume in the small bucket as $a : b$, the ratio in the large one will be $4\,a : 8\,b$. That is to say, there will be half as much surface exposed in proportion to the contents in the large bucket as in the small one. The heating and wasting of the air by evaporation is due to the surface exposed. Therefore, the larger the vessel, the less in proportion will the waste due to heating from the exposed surface be. If a bucket were five times as large, the ratio would be still more favorable—$25\,a : 125\,b$, or $1 : 5$, and so on.

By carrying out what the French would call the audacious idea of making liquid air by the barrelful, Tripler has demonstrated the possibility of handling it on the large scale pretty nearly as water is handled. The English scientists, as late as 1897, find it impossible to credit the accounts of what is done in this country. Prof. Fleming says that "nothing was effectual in storing liquid air until Prof. Dewar invented the silvered, vacuum-jacketed glass vessel as a container, and the even more effective and ingenious mercury vacuum process for introducing the high vacua required, without which none of our research work could have been done." This is not the only quotation which might be used to show how incredible the achievements on this side of the ocean seem to foreign investigators.

Liquid air, if it could only be produced cheaply enough, would represent an ideal substance for the production of energy. It is calculated that in one pound of it there are stored 139,100 foot pounds of energy. An electric storage battery varies from one-tenth to one-twentieth of this amount per pound of its own weight, and compressed air is about one-tenth. A pound of water compressed to 400 pounds pressure to the square inch has only one-quarter the energy of an equal weight of liquid air. In the compressed air and liquid air calculations the weight of the reservoir is not included.

The peculiarity of liquid air as a material for the storage of energy is that it can be made to give any pressure, from the slightest up to many atmospheres, nearly a thousand in number. It represents the water in a boiler, the containing vessel is the boiler, and the atmosphere represents the hot gases and flames of the furnace. By exposing more or less of the surface of the vessel to the air the evaporation could be controlled. Its expansion would tend to be adiabatic, but by further use of an air reheater, identical in construction with an air condenser, the disadvantageous adiabatic element may be suppressed, and isothermal, or nearly isothermal, expansion substituted. The condition is as if steam were superheated between boiler and engine, and as if the engine itself were heated by an external fire.

	Substance.	Symbol.	Critical temperatures. Degs. Cent.	Critical temperatures. Degs. Fahr.	Critical pressure. Atmos.	Temp. of saturated vapor at atmospheric pressure. Degs. Cent.	Temp. of saturated vapor at atmospheric pressure. Degs. Fahr.	Freezing point. Degs. Cent.	Freezing point. Degs. Fahr.	Pressure at which freezing point was determined. mm.	Density of gas.	Density of liquid at temperature given.	Color of liquid.
1	Water........	H_2O	365	689	200	100	212	0	32	760	....	1 at 0·4° C.	Colorless.
2	Hyd. selenide.	H_2Se	138	280·4	91	− 41	− 41·8	− 68	− 90·4	....	40	....	"
3	Ammonia	NH_4	130	266	115	− 33	− 27	− 77	− 107	....	8·5	0·6364 at 0° C.	"
4	Propane....	C_3H_8	97	206·6	44	− 45	− 49	Still liquid at −151° C.		....	20·95	....	"
5	Acetylene.....	C_2H_2	37	98·6	....	− 85	−121	− 81	− 113·8	950	12·97	....	"
6	Nitrous oxide	N_2O	35	96	75	− 89	−128	−115	−175	760	21·99	....	"
7	Ethane.	C_2H_6	34	93·2	50·2	− 93	−135·4	Still liquid at −151° C		....	19·97	....	"
8	Carb. dioxide	CO_2	31	88	75	− 80	−112	− 56	−69	760	21·94	0·83 at 0° C.	"
9	Ozone.........	O_3		....	....	−106	−158·8	....	...	...	23·89	....	Dark blue, easily exploded.
10	Ethylene.	C_2H_4	10	50	51·7	−102	−150	−169	−272	....	13·97	....	Colorless.
11	Methane......	CH_4	− 81·8	−115·2	54·9	−164	−263·4	−185·8	−302·4	80	7·98	0·415 at −164° C.	"
12	Nitric oxide ..	NO	− 93·5	−135	71·2	−153·6	−254	−167	−369	138	14·98	...	"
13	Oxygen.......	O_2	−118·8	−181·4	50·8	−181·4	−294·5	....	....	....	15·96	1·124 at −181·4° C	Blue.
14	Argon....	A	−121	−185·8	50·6	−187	−304·6	−189·6	−309·3	....	19·9	about 1·5 at −187° C.	Colorless.
15	Car. monoxide	CO	−139·5	−219·1	35·5	−190	−310	−207	−340·6	100	13·96	....	"
16	Air......	..	−140	−220	39	−191·4	−312·6	....	....	....	...	0·933 at −191·4° C.	Light blue.
17	Nitrogen......	N_2	−146	−231	35	−194·4	−318	−214	−353·2	60	14·01	0·885 at −194·4° C.	Colorless.
18	Hydrogen ...	H_2	−234	−389	20	−243	−405	....	....	....	1	....	"
19	Helium.......	He	....	....	....	Below − 264	−443·2	....	....	....	2·02	....	

Data collected and tabulated by Walter H. Dickerson, M.E.

CHAPTER XVIII.

CLAUDE AND LINDE.

Nitrogen of the Air and Its Utilization—Georges Claude—New Importance of Oxygen—Perfect Gases—Dewar's Cycles—Linde's Principle—Low Temperature Regeneration—Claude's Investigations and Their Developments—Gasoline, a Low Temperature Lubricant—Interchangers—Claude's Apparatus—Relations of Oxygen and Nitrogen in Liquefaction and Evaporation—Dew Point Production of Liquid Gases—Retrograde Condensation or Backward Return—Rectification—Claude's Oxygen Apparatus—Linde's Nitrogen Apparatus—Claude's Nitrogen Apparatus.

The liquefaction of air and of other gases may rank as one of the miracles of science. At first it appeared little more than a triumph of human intellect, for the practical element seemed lacking; now that we had liquid air and could carry it about like water, what was to be done with it? Pictet seized upon it as a basis for the production of oxygen. But the demand for oxygen was limited in those days. About one-fifth of the air is oxygen. In extracting from it its oxygen or rather a part of its oxygen, for some goes to waste, the other four-fifths were allowed to escape as a waste product, for no one wanted nitrogen gas.

As air contains so much nitrogen it is obvious that it costs about one-fourth as much to make a given volume of it as of oxygen. If a cubic foot of oxygen is produced, about four feet of nitrogen are simultaneously separated. Then came the development of processes for the synthesis of nitrogen products from the gas and hydrogen, either direct or indirect. This gave a new outlook to the liquid air industry. Ammonium salts are bases for numerous products used on an enormous scale in agriculture and in the manufacture of dyes, of explosives and of many other substances. Liquid air is now no longer a mere milestone on the road of human progress, it is a factor in the growth of crops, and in much that concerns the larger processes of human operations.

A new name appears in this chapter—it is Georges Claude. In the course of a number of years of patient work he has taken a place beside Linde and the other investigators who have developed the practical side of liquefaction of gases. But it must be remembered that it is the utilization of the nitrogen of the air that has made the subject a practical one, and that has converted it from a laboratory process to a great manufacturing industry. For oxygen is not needed to the same extent as is nitrogen. It even seems a dispensation of providence that nitrogen is the preponderating constituent of our atmosphere. And like Pictet and Linde, Claude developed his apparatus and processes for the production of oxygen; the all-important nitrogen was considered as of little or no importance. Hereafter it is to grow our crops, helping to feed the world, and to enter into explosives for less benevolent purposes.

Oxygen also is of new importance with the extension of its uses in industry, with acetylene in welding and cutting iron especially. The oxygen blow-pipe using blau gas or acetylene enables quartz to be worked by the glass blower. Autogenous soldering marks another application of the oxygen blow-pipe in industry.

A perfect gas is subject to Dalton's law, which may be thus expressed: The volume of a perfect gas varies directly with the pressure, so that the product of the pressure multiplied by the volume it occupies at that pressure is a constant; or pv is a constant for all values of p and v. The letters indicate respectively pressure and volume corresponding thereto.

As a gas approaches its liquefaction point it departs from this law and ceases to be a perfect gas. Down to a temperature of $-150°$ C. at atmospheric pressure, air is almost a perfect gas, but as the pressure is increased the product, pv, begins to diminish, the diminution at 15 atmospheres being considerable. At 29 atmospheres and $-145°$ C. it is a very little over one-half its value at atmospheric pressure, and at 40 atmospheres and $-140°$ C. it is less than one-quarter its value as a perfect gas.

Referring the question to temperatures we find that air acts as an almost perfect gas as follows:

At 0° C. (32° F.) up to 130 atmospheres;
At $-35°$ C. ($-31°$ F.) up to 45 atmospheres;
At $-130°$ C. ($-202°$ F.) up to 15 atmospheres;
At $-140°$ C. ($-220°$ F.) up to 10 atmospheres.

The reduction of the value of pv is attributed to the increased attraction of the molecules of the air for each other, owing to their greater proximity to each other, as compression brings them together.

This condition applies to other gases, with the exception, that in this aspect, hydrogen acts differently and as a more than perfect gas.

The application of this in the liquefaction of gases is important. The liquefaction processes are based on the disappearance of heat-energy as the gas under treatment develops mechanical energy in its expansion. Therefore the compression and expansion of the gas should be conducted within the limits of the range of temperature and of pressure within which it remains a perfect or nearly perfect gas.

Another way of putting the above is this. When a gas is near the liquefaction point, it has little power of developing cold by expansion, because it loses its expansion-power. Thus at —135° C. (— 211° F.) the loss in expansion-power is 100%; at —140° C. (—220° F.) it is 300%.

Dalton's law does not hold, as we have seen, for liquefactions. In the case of mixtures if the process is carried out at a constant temperature the pressure will vary and the composition of the liquid will vary. The liquid invariably contains some of the more difficultly liquefiable constituents. Thus a mixture of 6 parts of carbonic acid gas and 1 part of nitrogen begins to liquefy at 3·5° C. (38° F.) and 48·3 atmos., and the liquid will invariably contain some nitrogen, although the critical points of nitrogen are so far remote from those of the easily liquefied carbonic acid gas. The same obtains in other mixtures, the easily liquefied gas carrying some of the more refractory gas with it into the liquid state in solution.

Dewar's work has been largely dependent on the utilization of an outside source of cold. One of his cycles of operation, sometimes called his cascade, starts with carbon dioxide. This is liquefied by pressure at ordinary temperatures. It is gasified with the production of cold, which is used in the liquefaction of ethylene. The latter in its turn being gasified produces the working temperature of the apparatus. The gaseous carbon dioxide is liquefied as before, it acts in the liquefaction of the ethylene and so the cycle goes on continuously. The essential feature of the system is that it starts without reduction of temperature, because carbon dioxide liquefies at ordinary temperature by simple pressure.

The following relation of energy expended in compressing a gas and the pressures involved is sometimes called Linde's principle. The law is that the energy expended in compressing a gas from one pressure to another is proportional to Log. $\frac{p^1 \text{ (high pressure)}}{p^2 \text{ (low pressure)}}$ Substituting atmospheres of pressure, we get the following results for two cases:

$$\text{Log.} \frac{200}{1} \text{ atmos.} = 2{\cdot}30 \text{ and Log.} \frac{200}{50} \text{ atmos.} = 0{\cdot}60$$

which tells us that only a little over one-quarter the energy is required to compress air at an initial pressure of 50 atmospheres to a final pressure of 200 atmospheres as is required if the initial pressure is one atmosphere. But in practice the frigorific effect is nearly three-quarters as great in the application of the second case as in the first.

When air is liquefied the first sign of the change is the formation of a drop of liquid. The point at which this drop forms is called the dew-point, although it is not a formation of dew, properly speaking. This first drop is not pure air, it is a liquid containing forty-seven per cent. oxygen and fifty-three per cent. nitrogen. Air contains twenty-one per cent. oxygen.

There is a converse to this. It requires a forty-seven per cent. oxygen mixture in the liquid state to produce normal twenty-one per cent. air by its evaporation.

In liquefaction processes the gas which escapes liquefaction is reduced to about the temperature of the liquefied product. Suppose that this cold gas is made to give up its cold to the gas entering for treatment, an exchange of heat for cold will be effected, and the apparatus is called an interchanger. A good interchanger will bring the difference of temperatures between entering and leaving gases down to but three or four degrees C. One feature of the interchanger is that, when it is a part of an apparatus liquefying any gas, the escaping gas will necessarily be less in amount than the gas which enters. Therefore, the entering gas has its temperature reduced less than that of the escaping gas is raised.

The idea of regeneration of low temperature or of interchange now universally used in liquefying apparatus, as we have seen, goes back many years, being described by William Siemens in a patent in 1857. The cylinder and piston idea goes back to 1898, although Claude was the first to successfully apply it several years later, only succeeding in overcoming the difficulties of lubrication after some years of work, during which he was developing the process, and has developed the practical aspect in his very exhaustive work on the liquefaction of gases and the production of oxygen, hydrogen, helium, nitrogen and other elements.

Claude's work in the attempt to produce oxygen from the air began in 1899, when Linde had perfected his apparatus. Claude tried to separate oxygen and nitrogen by centrifugal action, as cream is separated from milk. He also tried a solution process with alcohol as the solvent to enrich the air by dissolving the nitrogen.

A year earlier Lord Rayleigh had calculated that if the expansion of air in Linde's apparatus were made to drive a turbine, the result would be greatly improved. It has been found that air escaping through a diaphragm or valve and falling from a pressure of 40 atmos. to 1 atmos. experiences a reduction of temperature of only 10° C. But if such air operates against a piston, thereby exerting mechanical energy, its temperature will fall over 125° C. In this lies the principle of Claude's apparatus. Air under high pressure is admitted behind the piston of an engine of the ordinary reciprocating type. On its way to the engine it passes through an interchanger and is greatly reduced in temperature. As it leaves the engine, in whose cylinder it has expanded, it liquefies in part, and is collected in a tank or receptacle. Part goes off in the gaseous condition through the interchanger, cooling the entering air.

Thus Claude's machine reduces in conception to a compressed air expansion motor operated in conjunction with an interchanger. The piston is driven a part of its course by direct pressure. During this phase no refrigeration takes place. The inlet valve closes at an early part of the stroke, operating as a cut-off, expansion takes place in the cylinder, driving the piston and expending mechanical energy, refrigeration begins and the temperature falls correspondingly.

The lubrication of his engine at low temperatures was an early problem. Claude found that gasoline operated perfectly. Once a good amount of liquid air was produced the liquid itself acted as a lubricant, and gasoline could be dispensed with.

A great improvement was effected by the use of leather for the packing of the piston instead of iron piston rings. Leather softens in intense cold, a very curious property and quite exceptional. It needs no

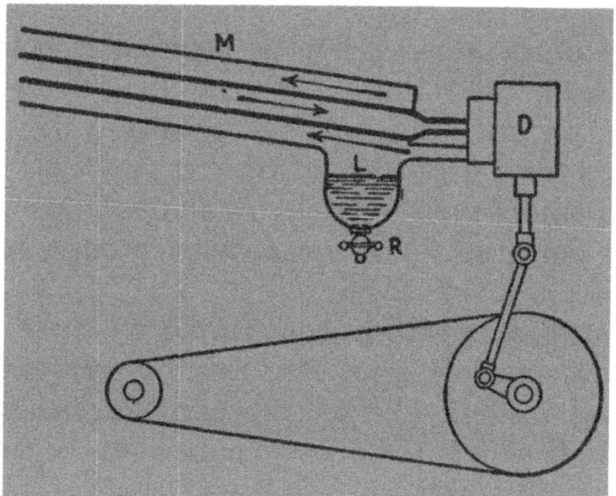

Principle of Claude's Apparatus.

lubrication whatever, and works far better than the regular iron piston rings.

In a single expansion Claude gets an 80 per cent. oxygen mixture and 99·8 per cent. nitrogen, which is practically pure. This purity is particularly important in the cyanimide process, to avoid waste of the carbide.

The cut illustrates in diagram the principle of the Claude apparatus. To the left is the interchanger, *M*, indicated by two concentric tubes. Through the outer one cold gas from the apparatus passes, getting warmer as it imparts its cold to the entering air or other gas which passes through the central tube. It would be more correct to say that it abstracts the heat from the entering gas. A part of the gas which enters is liquefied, so that the outgoing stream is less in amount than the entering one. At the outer end of the interchanger, the left end in the cut, there should not be more than three or four degrees difference of temperature between the two streams of gas.

The air or other gas to be liquefied is pumped in through the central tube. It enters the cylinder of the engine, on the right, and by a cut-off valve system, expands for a great part of the stroke. This cools it. It then enters the receptacle, *L*, and as it rushes in, a part is liquefied. The rest escapes through the outer tube of the interchanger.

Compound apparatus constructed on these general lines and on these principles, may have several cylinders and pistons, like a compound steam-engine, but the Claude system is fully summarized in the simplified diagram given here.

The power expended on driving the air-engine is not wasted altogether, as it is used as a motor to drive a dynamo or for other service.

The general principle of Claude's apparatus having been understood, the next cut shows the system of applying it to the liquefaction of gases. The air, assuming it to be a liquid air machine, enters through the central tube of the interchanger, *M*, is cooled by the gases escaping through the outside division of the same, and part of it enters the air-engine, *D*, actuating it and being greatly reduced in temperature. Part of the entering air, as shown in the diagram, goes into the top of the liquefier, *L*, and surrounds the tubes, cooling them to —140° C. The air exhaust from the engine is liquefied inside of the tubes of the liquefier in part, and part escapes through the outer tube of the interchanger. The characteristic feature of the apparatus is the division of the entering air, cooled and

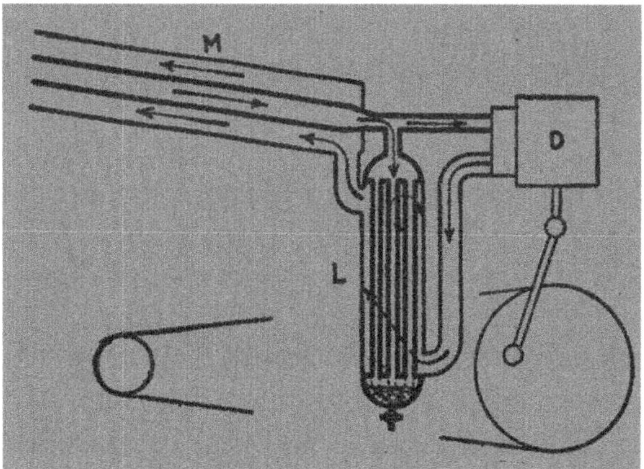

Claude's Compound Liquefaction Apparatus.

compressed to 40 atmospheres, between the air-engine and the liquefier.

Nitrogen evaporates from the liquid state quicker than oxygen. This fact was the basis of the first methods of making oxygen from the air. On exposure to evaporation liquid air gives off its nitrogen much more rapidly than its oxygen, the residual mixture grows richer in oxygen, and by successive evaporations or analogous methods a nearly pure oxygen is obtained. The nitrogen in these early processes was allowed to escape as being of presumably no value, but now it is the nitrogen which is sought for the manufacture of ammonium salts more than the oxygen.

As regards the condensation of the two gases in the liquefaction of air, various opinions were maintained by different authorities. Dewar and Linde considered that both condensed together, Pictet thought that nitrogen condensed first. If both gases liquefy together, preserving the percentages in which they exist in air, then it follows that to separate the two from each other as they exist in air, the first step must be to completely liquefy the air and to depend on the evaporation of the liquid to separate them. This proved not to be the case, and Claude in his early investigations found that the first liquid produced in the liquefaction of air was richer in oxygen than was the original air, and that the escaping or residual gas as a necessary sequence was far richer in nitrogen. It only needed a proper manipulation or distribution of the process to produce a practically pure nitrogen gas. He found that pure nitrogen could be obtained by liquefying only one-half of the air.

The modern processes for producing nitrogen and oxygen are based on the laws of the liquefaction of mixed gases as they apply to air.

The temperature of liquid nitrogen evaporating at a pressure of 0·07 atmosphere, almost a complete vacuum, falls to −213° C. (−351° F.) under which conditions the nitrogen in air will freeze while the oxygen will remain liquid, for curiously enough oxygen does not solidify at this temperature. A sort of sponge of solid nitrogen charged with liquid oxygen is produced, whence the liquid oxygen can be to some extent squeezed out as water is squeezed out of a sponge.

This is a way of separating the two gases, of interest from an experimental standpoint only.

Claude, in the year 1903, found that if air is exposed to a liquefying condition, and if the liquefied part is withdrawn from contact with the gaseous part as fast as it is formed, the liquid will contain as high as 48 per cent. of oxygen. This is in line with the so-called dew-point product of a 47 per cent. liquid oxygen. Here he found the clue to the desired way of separating the oxygen from air; he had no idea that soon it was to be the nitrogen which was desired. The gaseous portion escaping from the apparatus was rich in nitrogen, and in those days was considered a waste product.

Suppose the air compressed and at low temperature to be admitted to the base of the apparatus and to be partly liquefied there, the gas rising and ascending through the apparatus would be rich in nitrogen. Now assume that liquefying conditions are maintained in the apparatus above the liquid first formed, and that the liquid is drawn off as formed, then a constant liquefaction will take place in the rising stream of gas, and it will be constantly robbed of its oxygen. If the supplementary liquefying process be carried far enough, the escaping gas will be practically pure nitrogen. The condition described of gas rising through a cooling apparatus, with partial liquefication, is termed retrograde condensation.

Of course, retrograde condensation could be made to liquefy all of the gas, but it is generally understood to apply to this partial process only, bringing about the separation of air into its constituents. It is a species of rectification.

Retrograde condensation is also called backward return, rectification and counterflow.

The counterflow or backward return system of making nitrogen gas is based on the fact that oxygen liquefies at a higher temperature than does nitrogen. If air is liquefied by cold and pressure the first drop which forms contains 47 per cent. of oxygen, more than twice the percentage of that in normal air. Suppose that a stream of air under liquefying conditions entered a vertical tube and rose through it, liquefying as it did so. The first liquid formed would be rich in oxygen. This is the liquid at the bottom of the tube. As the impoverished air rose and liquefied it would form a constantly impoverished liquid, one poorer and poorer in oxygen. The rising air, already rich in nitrogen and poor in oxygen, would be subjected as it rose to the action of the liquid products descending through the tube. These would absorb any oxygen in the rising gases and it would be only a question of making the tube long enough to get a practically pure nitrogen. The rising gas would be constantly poorer in oxygen, and the liquid acting on it would be constantly poorer in oxygen also. This would give it greater absorbing power just when it was needed, for it is obvious that the gas poor in oxygen would need a liquid of high absorbing power, or what is the same thing, one poor in oxygen, to take out the last traces of oxygen from it. The principle is also termed rectification, and can be applied to mixtures of other gases.

Linde's apparatus for the production of pure oxygen employs the principle of rectification. The machine is shown in diagram in the cut, and the course of the air and constituent gases through the machine may be followed there. *G* is a rectifying column, filled with baffle plates or some similar device to break up the flow of liquefied gases through it, so as to bring

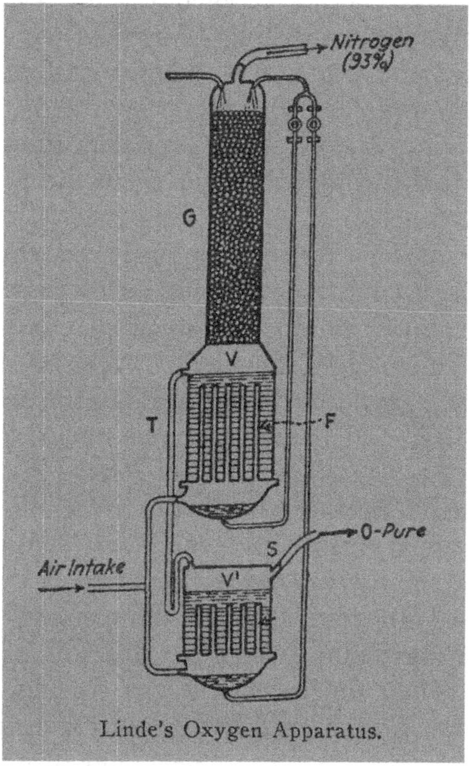

Linde's Oxygen Apparatus.

about a good contact between the ascending gases and the descending liquids. The vessel, *V*, is first of all filled with pure liquid oxygen. This surrounds the tubes as shown, it vaporizes and thereby cools them still more, and rises as a gas through the rectifying column, *G*. Air is pumped in under pressure at the intake and is liquefied in the tubes, *F*, in the vessel, *V*, pours down and collects in the receptacle below the vessel, *V*. Thence it rises to the top of the column, *G*, runs down over the baffle plates, losing nitrogen as it descends, and taking oxygen out of the ascending gases, and finally reaching *V* as nearly pure oxygen. It replaces the evaporated oxygen and adds oxygen to more than replace what has evaporated. The excess overflows into the vessel, *V'*, reenacts the cooling rôle and goes off through the pipe, *S*, to the gasometer or other receptacle. It will be seen that *V'* supplements the work of *V*. As the oxygen leaves the apparatus through *S*, it cools the incoming air by passing through an interchanger as already described. Ninety-six per cent. to ninety-eight per cent. oxygen is produced in this machine, and about two-thirds of the oxygen of the air is collected. The nitrogen is about 93 per cent. pure.

The next diagram shows the principle of Linde's nitrogen apparatus. *C* is a rectifying column. Above it is a coil, *S*, surrounded by liquid air, *e*. Cooled and compressed air rises through the column, *C*, meeting liquid air descending from *S*, where it has been liquefied. When the ascending gas reaches the top of the column, *C*, it is nearly pure nitrogen, the oxygen going down as a liquid. The liquid produced in the coil, *S*, is now liquid nitrogen of very great rectifying power, so that the nitrogen is produced purer than at first. The vessel, *e*, is constantly replenished from the compartment, *F*, through the cock, *g*. Nitrogen from the top of the column, *C*, is liquefied in the coil, *d*, in the vessel, *h*, in which vessel liquid oxygen from the rectification column, C_1 C_2, highly rectified, and cooled by its contact with liquid nitrogen collects and surrounds the coil. Forty-seven per cent. oxygen goes

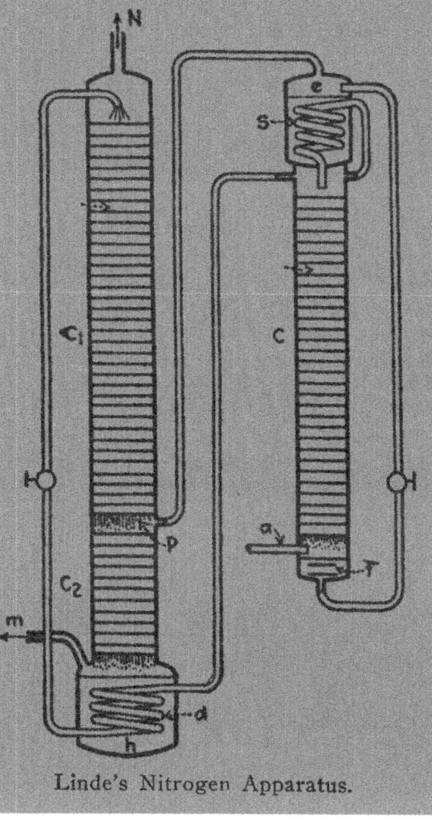

Linde's Nitrogen Apparatus.

from the first or right hand division of the apparatus into the second rectifying column, C_1 C_2, at the point, p, liquid nitrogen from the coil, d^1 rises through i, and enters the top of the column C_1 C_2. Part descends through the column, none reaching the bottom, and the rectification gives a very pure nitrogen escaping at *N*, and going to the gasometer or other apparatus and

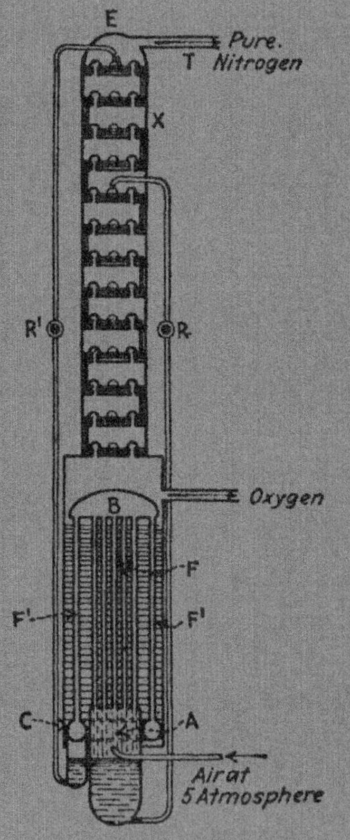

Claude's Nitrogen Apparatus.

pure nitrogen is collected at *N*, and oxygen is collected from *m*.

Claude's nitrogen apparatus is the subject of the

next diagram. Air, under pressure of 5 atmospheres and cooled to a low temperature in an interchanger, enters the division, *F*, at *A*. The tubes are cooled by liquid air surrounding them as indicated. Rising through *F*, the air is partly liquefied and some of it collects in *C*, as about a 47 per cent. oxygen. This is oxygen of what is called, although incorrectly, the "dew-point." The part which escapes liquefaction is approximately pure nitrogen. The nitrogen descends through *F*, is liquefied in part, and as such collects in *C*. The 47 per cent. oxygen is admitted in part into the rectifying column, *X*, flowing through the pipe, *R*. This is a beginning of the rectification. To complete it liquid nitrogen passes through *R'*, into the top of the rectifying column, completing the rectification, so that almost pure nitrogen passes out at *T'* to be saved, while oxygen is collected on a lower level, as shown.

CHAPTER XIX.

UTILIZATION OF ATMOSPHERIC GASES.

Cyanamid Process for Utilization of Atmospheric Nitrogen—Calcium Carbide as Used in the Cyanamid Process—Direct Use of Cyanamid as a Fertilizer—Cyanid Process for the Utilization of Atmospheric Nitrogen—Heber's Process for the Same—Literature of the Nitrogen Industries.

Cyanamid is produced by acting on calcium carbide with nitrogen. Calcium carbide is made by the reaction of calcium oxide (lime) on carbon in an electric furnace at an enormously high heat. The reaction is as follows:

$$Ca\,O + 3\,C = Ca\,C_2 + C.$$

$Ca\,C_2$ is the formula of calcium carbide.

When this is acted on by nitrogen the following reaction takes place:

$$Ca\,C_2 + 2\,N = Ca\,C\,N_2 + C.$$

$Ca\,C\,N_2$ is the formula of cyanamid.

When cyanamid is reacted on by water, ammoniacal gas ($N\,H_3$) is evolved by the following reaction:

$$Ca\,C\,N_2 + 3\,H_2\,O = 2\,N\,H_3 + Ca\,C\,O_3.$$

All salts of ammonium can be made with the ammoniacal gas thus obtained. The process is the synthesis of nitrogen and hydrogen by purely chemical reaction. The above reaction is exothermic or productive of heat.

The calcium carbide is finely ground out of contact with air, and is heated to redness in the presence of nitrogen. After the reaction is completed, acetylene, of which a certain amount is always present, is removed by exhaustion to a partial vacuum, and the cyanamid, mixed with calcium carbonate and lime, is treated with steam, and the reaction last given takes place, the calcium carbonate and lime not figuring in it as active members.

The cyanamid reaction, the one given above in the second place, is carried out in autoclaves. These are heavy iron vessels hermetically sealed, their lids being held in place by powerful screw fastenings. The reaction starts at 3 to 4 atmospheres and in twenty minutes the pressure rises to 12 or 15 atmospheres. After this it falls. The autoclaves are provided with relief valves to prevent explosions.

Cyanamid is so dusty or pulverulent that it cannot be conveniently used as a fertilizer, although it is otherwise adapted for direct use. The difficulty has been overcome by mixing it with 15 per cent. of tar.

In the cyanide process, carbon, and soda ash (sodium carbonate) are heated together in a stream or atmosphere of nitrogen. The sodium and a part of the carbon and the nitrogen unite to form sodium cyanide, and the rest of the carbon unites with the oxygen to form carbon monoxide. The reaction is as follows:

$$Na_2CO_3 + 4\,C + 2\,N = 2\,NaCN + 3\,CO.$$

In carrying out the process, the soda ash is mixed with finely ground coke or other form of carbon and with finely divided iron. The whole is heated to redness, 1,000° C. (1,832° F.) and nitrogen is passed through it. From the cyanide thus obtained ammoniacal salts and nitric acid can be readily obtained. The iron acts as a catalyzer, taking no part in the reaction.

The sodium cyanide, produced as above, can be treated with steam at a low temperature, producing sodium formiate and ammoniacal gas, thus:

$$Na\ C\ N + 2\ H_2\ O = Na\ H\ C\ O_2 + N\ H_3.$$

On heating sodium formiate the original amount of sodium carbonate is recovered:

$$2\ (Na\ H\ C\ O_2) = Na_2\ C\ O_3 + C\ O + H.$$

Ammoniacal gas, $N\ H_3$, is an exceedingly available form for the utilization of nitrogen in ammonium compounds.

In the Haber process, nitrogen and hydrogen gases are heated together in the presence of finely divided iron at a temperature exceeding 500° C. (932° F.) and at a pressure of 125 to 150 atmospheres. This process has been quite extensively used. In it as in the cyanide process the iron plays only the inactive rôle of a catalytic, not entering into the reaction, which is a simple and direct combination of the two gases to form ammoniacal gas, by the reaction:

$$N + 3\ H = N\ H_3.$$

An excellent article on the literature of the nitrogen industries by Helen R. Hosmer, of the General Electric Company, may be called to the reader's attention. It is published in the *Journal of Industrial Engineering Chemistry,* vol. ix, No. 4.

CHAPTER XX.

Helium, Argon and Various Notes.

Charcoal Absorption of Gases at Low Temperatures—Charcoal of Highest Absorbing Power for Gases—Kammerlingh-Onnes' Liquefaction of Helium—The Apparatus Used—Use of Helium for Balloons—Use of Argon for Incandescent Lamps—Liquid Hydrogen Vacuum—Variation of Liquefaction Temperatures of Air—Chemical Affinities at Low Temperatures—Liquid Ozone—Blasting Cartridges from Liquid Air.

Hydrogen has been made in quantity by both Linde and Claude by liquefying the carbonic monoxide in water-gas, the hydrogen passing on as a gas. One of its uses was to be the preparation of synthetic ammonia, $N\ H_3$, as in the Heber process summarized in the last chapter.

It has been found that charcoal at low temperatures possesses very strong absorbing power for air. Fifteen grams, or about half an ounce of charcoal, at what may be termed liquid air temperatures, will absorb the contents of a liter and a half flask, nearly three quarts, no less than twenty-five times. The charcoal vacuum up to a certain extent will exceed that of a Crookes tube. The illustration shows a copper container for liquid gases, in whose outer space the vacuum is maintained by a piece of charcoal contained in the inclined tube. It answers every purpose of a

Dewar flask, and operates on exactly the same principle. The charcoal at the low temperature produced by the liquid gas in the inner compartment, absorbs the air in the outer compartment completely. Such a flask has held liquid air for a week. Claude saw the possibility of making large containers on this principle, which would be free from danger of breaking

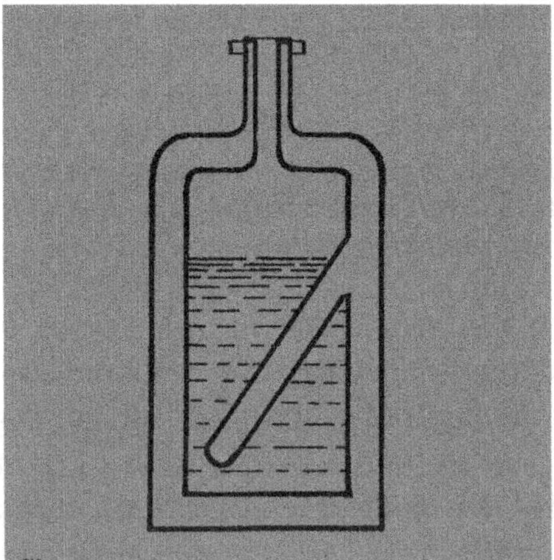

Charcoal Vacuum Liquid Gas Container.

and could be used of large sizes, for commercial transportation of liquid gases, but a patent of Dewar's interfered with the development of this idea.

The best kind of absorptive charcoal is made from cocoanut shells. It should be recently prepared and should be heated to 400° C. (752° F.) to free it from all gases before using it in low temperature absorptions of gases. One gram will absorb at low temperatures from 300 to 400 cubic meters of air. At

0° C. (32° F.), it will retain two volumes of helium and four of hydrogen. This is a partial separation, but at −185° C. (−301° F.) the proportion is much

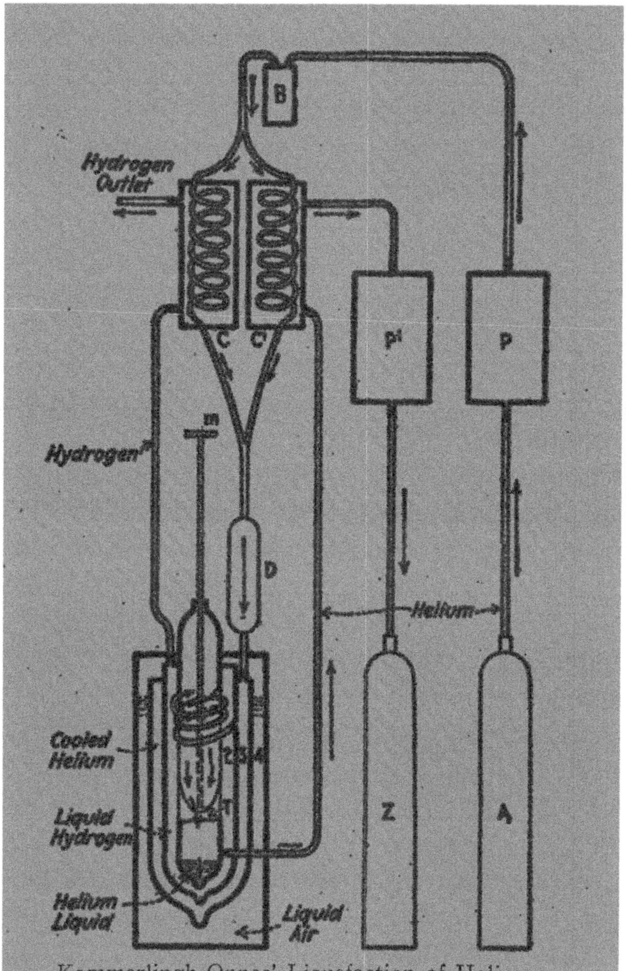

Kammerlingh-Onnes' Liquefaction of Helium.

more favorable. At this temperature one volume will retain 135 volumes of hydrogen and only 15 of helium. Impure helium can be purified from other gases by well known absorbents, such as sodium or calcium hydroxide for carbon dioxide, red-hot oxide of copper for nitrogen and so on. As helium is almost without chemical affinities, the purification of it is comparatively easy. The charcoal purification removes the hydrogen. Kammerlingh-Onnes succeeded in liquefying it in 1908 in the cryogenic laboratory at Leyden, Holland. He subjected it to a temperature of —258° C. (—432° F.) at a pressure of 100 atmospheres. He assigned it a density of 0.154, boiling point of —268° C. (—450° F.), and could not solidify it.

The general outlines of the Kammerlingh-Onnes apparatus are shown in the cut. The purified helium is contained in the receptacle *A*. It is compressed by the pump, *P*, and passes through a desiccating vessel, *B*, cooled by liquid air. It then passes through two worms. One of these is cooled by hydrogen and the other by helium, both gases escaping from the liquefied elements, as will be seen. The helium now passes through charcoal contained in a purifier, *D*, where it is freed from any hydrogen it may be mixed with. It is still further cooled in the vessel, 3, which is immersed in liquid air, the latter in the exterior vessel, 4. From 3 it circulates through a coil in the vessel 2, where it is exposed to the gas from boiling hydrogen and enters the central vessel 1. It escapes therefrom through the stopcock *T*, operated by the handle *m*. A part is liquefied and collects as indicated. The gaseous helium from the liquid cools the worm *C'*; the other worm is cooled by the hydrogen gas from the vessel 2.

The escaping helium is pumped into the receptacle *Z* by the pump *P'*.

According to Claude, there are 5 volumes of helium in 1,000 volumes of air.

On page 276 of the present book a remark of Professor Dewar is cited to the effect that a regular gas liquefaction apparatus could be installed at Bath, England, and made to produce any quantity of helium, if there were any demand for it. Helium is a gas of about twice the specific gravity of hydrogen, and is of very slight affinity for any other element. It is not combustible. It is now proposed to use it for balloons.

The inflammability of hydrogen has operated to greatly increase the danger in the use of balloons of the self-propelling type, such as Zeppelins. Gas bags filled with thousands of cubic feet of hydrogen are very uncomfortable neighbors of a group of gas engines directly below them, as there is always danger of the exhaust gases from the engines igniting the gas and causing a disastrous accident. In war the inflammability of the hydrogen-inflated dirigible made it a target for incendiary bombs. A helium-inflated balloon is free from these troubles. In times of peace it would be a far preferable and safer vehicle than the hydrogen balloon. It would admit of various modifications in construction. Thus it has been suggested that the engines might be placed within the casing of the balloon, if the latter were of the rigid type.

One thousand cubic feet of commercial hydrogen has a lifting power of seventy pounds. The same quantity of helium can lift sixty-five pounds.

In the Southwest plants of the two types, Linde and Claude, have been installed for the manufacture of helium from natural gas, which may have as much as 2% of helium. In the Linde plant the first product was only 67% pure; it was repurified to 92% to 93%, and was then available for balloons. During the Great War, work on its production was in active progress, and the operations were camouflaged under the pretence that argon was the gas experimented with or operated on.

Of the rare gases of the atmosphere, argon is used as a gas for the filling of incandescent lamp bulbs, and liquefaction processes produce it in commercial quantities.

Liquid hydrogen weighs only 100 grams to the liter; this is one-tenth the weight of water. Liquid argon has a specific gravity of about 1·50, or 1,500 grams to the liter.

The vacuum in a Dewar flask or other vessel can be produced by liquid hydrogen. A glass bulb is connected by a glass tube to the vessel to be exhausted. The bulb communicates with the vessel by the tube. The bulb is now immersed in liquid hydrogen. The air in the bulb liquefies and freezes; air from the vessel replaces it and in a very short time there is an almost perfect vacuum in the vessel. The tube is then melted off and sealed with the blow-pipe and the operation is complete.

An increase of temperature of liquefaction of air of about 8° C. is caused by each atmosphere increase of temperature.

Claude once placed at Dewar's disposal the residual gases from 100 tons of air. Dewar could find no new element in the residue other than those we have cited.

As a rule, chemical affinities are reduced to almost

zero value at low temperatures. There is one exception. Solid fluorine combines with liquid hydrogen at —250 C. (—418° F.) with explosive violence.

Ozone in the liquid state is indigo blue in color.

Cartridges for blasting have been made by pouring liquid oxygen on powdered charcoal; these can be exploded with a detonator. They have to be used as made. This is considered an advantage, as if they fail to explode there is no danger of an explosion, days or hours after the time when the explosion should have taken place. This sometimes happens with ordinary explosives, but the liquid oxygen cartridge is safe after a half hour.

SUPPLEMENTARY INDEX

PAGE

Dalton's law 360
Hydrogen, a more than perfect gas 360
Cascade, Dewar's 361
Linde's principle 362
Interchanger 363
Claude's work 363–366
Leather for packing 365
Claude's principle 365
Claude's apparatus 365–367
Nitrogen, freezing more readily than oxygen 368–369
Claude's discovery in the separation of nitrogen 368
Dew-point, so called 369
Condensation, retrograde 369
Backward return 370
Counterflow 370
Rectification 370
Linde's oxygen apparatus...370–372
Linde's nitrogen apparatus..373–374
Claude's nitrogen apparatus.374–375

PAGE

Cyanamid 376
Nitrogen, fixation of 376–378
Cyanamid as a fertilizer 377
Cyanide process for fixation of nitrogen 377–378
Haber's process 378
Hydrogen absorbed by charcoal 379–381
Claude's charcoal vacuum container 379–380
Kammerlingh-Onnes' liquefaction of helium 381–382
Helium, liquefaction of 381–382
Helium, use of, for balloons... 383
Helium, its lifting power in balloons 383
Helium camouflaged as argon.. 384
Helium, purity of 384
Argon, specific gravity of liquid 384
Fluorine at low temperatures.. 385
Blasting with liquid air 385

INDEX.

Absolute cold.................. .. 19-20
Absolute zero. 40-41
Acetylene, Cailletet's work on, 175, 179-181, 182-183
Adiabatic expansion and contraction.... 69
Air, a conveyor of heat 244
Air and water contrasted 85-86
Air, Cailletet's liquefaction of..186-187
Air, composition of 87
Air, constancy of composition of .89-90
Air, dry and wet compared ... 14-15
Air, experiments with liquid...325-337
Air, how to preserve liquid indefinitely............. . . 260-261
Air, liquefied, giving two liquids 220
Air, liquid, defined............. 9
Air of atmosphere not a chemical compound 86-87
Air, physicists' and chemists' views of 88-89
Air, Wroblewski's experiment on liquefaction of....... 220
Alcohol frozen by liquid air.... ...330
Alcohol frozen by Wroblewski and Olszewski 212
Amagat 147
Ammoniacal gas, Faraday's liquefaction of 111-112
Ampere and Colladon, anecdote of 133-135
Andreef 214
Andrews, memoir on life of, by Tait and Brown.... 150
Andrews, Thomas. 19, 133, 147-150, 169, 176
Apparatus and experiments, Cailletet's liquefaction......... 177-182
Apparatus and process, Tripler's 290-295
Apparatus, Hampson's, for liquefying air.............. ... 320-324
Apparatus, Linde's, for liquefying air....... 307-319
Apparatus, Pictet's liquefaction. 157-163
Apparatus, Thilorier's......... 137-141
Argon 87
Arseniureted hydrogen, liquefaction of............. 122
Atmosphere, its relation to animals and birds...... 85-86
Atmosphere liquefied 336
Autophage state of human system...............................339
Babbage 119-120
Barker, George F..............199, 240
Barleycorn as unit of space...... 25-26
Bath well, helium from........275-276
Battle of squares and cubes....355-356
Baucalari 115
Benjamin Thompson 93
Bianchi's modification of Natterer's apparatus..... 142-145
Blenkroode.... ... .245
Blenkroode's experiment illustrating vacuum.... 245
Boiling a cooling process.........76-77
Boiling by producing a vacuum..77-78
Boiling gases..................... 76-77
Bonty ... .201
Brunel's carbon dioxide engine 99
Buckets, Tripler's, for liquid air 289-296
Bulbs, efficiency of different.......247
Bulbs, vacuum, mercury silvering of..... ....247, 253-254
Cagniard de la Tour 128-133
Cailletet, L. P...22, 24, 58, 135, 150, 151, 155, 156, 165, 172-202, 204, 212, 214, 215, 218, 219, 220, 226

Cailletet and Hauteville on specific gravity of oxygen..........197
Cailletet, honors received by 174
Cailletet, life of................173–174
Cailletet, liquid acetylene, his work on........ 197–199
Cailletet on conductivity of metals at low temperatures 201
Cailletet on critical state phenomena 190–192
Cailletet performs La Tour's experiment.......................202
Cailletet's cold blast blowpipe, 141, 198–199
Cailletet's continuous liquefaction process....................200
Cailletet's control experiment with hydrogen 184
Cailletet's controversies with Dewar................. 232–233
Cailletet's frozen mercury stopper.... 187
Cailletet's letter to Academy of Science 183–184
Cailletet's liquefaction of hydrogen 218
Cailletet's manometers 187–189
Cailletet's thermometric methods, trials of....................201–202
Callendar...... 57
Carbon bisulphide, frozen by Wroblewski and Olszewski....212
Carbon burned in liquid air........334
Carbon dioxide in air.... 90–91
Carbon dioxide in liquid air........336
Carbon dioxide, liquefaction of, Faraday's.................... ..111
Carbon dioxide, solid 15–16
Carbon monoxide dispatch, Wroblewski and Olszewski's........213
Carnot's cycle 70, 288
Celsius thermometer scale........38–39
Centimeter........... ... .25
Chemical reactions governed by cold.................... ...347–348
Chlorine, Faraday's liquefaction of............. ....106, 110
Chlorine, Northmore's liquefaction of...... ....106, 117–118
Clark, Dr. G. Fish................ 342
Clausius........ 204
Coal as a chemical.................298
Cold, absolute 19–20
Cold, distillation by 347
Cold, regeneration of..........·.....299
Cold regenerative process.. 265
Coleman.,...... 265
Colladon and Ampere, anecdote of...........................133–135
Colladon, Daniel..133–137, 174, 176, 179, 200, 207
Colladon, his original apparatus...136
Conservation of energy...........29–36
Contraction, adiabatic..............69
Cordes, Dr ... .341
Count Rumford..................92–95
Critical pressure........ 19
Critical state of matter...........19–20
Critical temperature........ 19
Crookes layer.... 80–82
Crookes layer, protection due to....84
Crookes, William 284
Cubes and squares, battle of...355–356
Cyanogen, liquefaction of, Faraday's ... .111
Cycle of reversible engine 70
Davy-Faraday Research Laboratory 115
Davy, Sir Humphry.. ...96–99, 102, 103, 105, 110, 115, 120 121, 126
Debray........ ....211–212, 218
Dewar and Moissan's liquefaction of fluorine 276–280
Dewar, James..96, 99, 112, 115, 151, 157, 168, 198, 200, 206, 215, 219, 225, 227, 229, 230–285
Dewar's apparatus of 1883.... 233
Dewar's apparatus of 1895......238–239
Dewar's bulbs 244–254
Dewar's bulbs, mercury silvering of......................247, 253–254
Dewar's colleagues.... ... .232
Dewar's controversies with Cailletet.............. 232–233
Dewar's early apparatus.......233–237
Dewar's gas jet experiments...264–266
Dewar's hydrogen jet experiments.... 266–271
Dewar's life 231–232
Dewar's liquefaction of helium....281
Dewar's liquefaction of hydrogen.............. 280–285

Dewar's separation of helium..275-276
Dewar's small gas liquefaction apparatus 241-243
Dewar's suggestion of marsh gas as a refrigerant 232-233
Dewar's use of Pictet's cycles ... .233
Dewar's vacuum...... 249-253
Diffusion 18
Dog killed by low temperature ... 340
Double and triple glass gas bulbs.......................245 247
Ducretet.................. 206
Dufour, Prof Henri............155-156
Edison 288
Effects of intense cold on human system 339-341
Eiffel Tower manometer .. .188-189
Elasticity of metals affected by cold.... .. .. 255-256, 260
Electric power transmission, liquid air in........ 353-354
Electric resistance of metals affected by cold, Wroblewski on.... 219-220
Electrolysis of water 148-149
Elements, fundamental in physics .25
Elongation of metals affected by cold... 259-260
Energy ... .29
Energy and force 24-25
Energy, conservation of 29 36
Energy converted into useless heat...................... 29
Energy, kinetic 31
Energy, low grade heat........288-289
Energy, low grade heat and liquid air 35-36
Energy, potential 30
Energy, reduction of available...71-72
Energy, reservoir of. 32-33
Energy, unutilizable of world. ..34-35
Energy, waste of, in railroads and steam navigation 33-34
Entropy.................... 34
Ethylene.... 197-199
Ethylene, liquid as refrigerant.197-198
Ethylene, Wroblewski and Olszewski's results with. 212
Euchlorine, Faraday's liquefaction of.... 111
Evaporation by stream of gas. 201, 214
Expansion, isothermal...........68-69
Experiment, Blenkroode's, showing utility of vacuum . .. 245
Experiment, Count Rumford's.118-119
Experiment illustrating conservation of energy................32
Experiment in boiling by a vacuum 77
Experiment, Joule and Thomson's.... 61-62
Experiment, Joule's..... 60-61
Experiment on low grade heat energy........ 35-36
Experiment with chlorine hydrate 126
Experiment with india rubber band.... 32
Experiment, Villard's. 24
Experiments, Dewar's hydrogen jet..................... 266-271
Experiments, Dewar's, on solutions of gases in other liquefied gases.... 271-274
Experiments, Dewar's, with gas jets 264-266
Experiments, early, of Faraday...103
Experiments in spheroidal state.82-83
Experiments, La Tour's..... .129-132
Experiments, Pictet's, of 1877..160-161
Experiments with liquid air...325-337
Explosions in Faraday's and Davy's early work..............110
Explosive, liquid air..... 348-353
Faraday as fellow of the Royal Society 106-107
Faraday, Michael..... 28, 42, 95, 99, 100-115, 117-129, 131, 226, 240
Faraday, Michael, his life 100 115
Faraday on Davy's continental tour.................. 105
Faraday's bent tubes 123-128
Faraday's death 115
Faraday's discovery of magnetism of oxygen...... 114-115
Faraday's engagement at Royal Institution............... .104-105
Faraday's failures in liquefactions of gases 114
Faraday's liquefactions of gases, 106-112, 113-114
Faraday's solidification of gases ..114

Faraday's thermometer........ 39
Fleming, J. A 232
Fluorine, liquefaction of...... 276-280
Force 27-29
Force and energy.............. 24-25
Force, conservation of, an erroneous doctrine.... 28-29
Force, living.......... 29
Formula, Joule-Thomson effect, 300-306
Frigotherapy.......... 338-342
Fuller, Mr. John 95
Fullerian professorship in Royal Institution.. ....95, 96
Gas cooled by expansion 299
Gaseous state of matter.......... . 12
Gases, boiling........ 76-77
Gases, Davy's experiments in inhaling............... 97-98
Gases, Davy's views of the utility of liquefying............... 98
Gases, determining latent heat of liquefied....................261 264
Gases, determining specific heat of liquefied.................261 264
Gases, molecular motion in......17-18
Gases, permanent 149-150
Gases, solution in other liquefied gases 271-274
Gas heavier than liquid 21
Gas jets, Dewar's experiments with 264-266
Gas, receiver for liquefied, Cailletet's 195-196
Gas, the perfect... 59-62
Galbanum..........................191
Galitzine 22
Gramme..................... 25
Griffiths..........................57
Hampson.....226, 238, 265, 300, 301, 309, 320-324
Hannay. 23
Hauteville....... 197
Heat, latent. 72-76
Heat, measurement of... 37
Heat of ice.......................11
Heat, specific. See specific heat.
Heat, utilization of unavailable .. .72
Helium, liquefaction of Dewar's. 281
Helium, separation of, Dewar's, 275-276
Helmholtz......... 204
Hervy, death of 138
Hogarth.... 23
Hydrochloric acid, Faraday's liquefaction of 112
Hydrogen, Cailletet's liquefaction of 184-185
Hydrogen, constants of liquid by Olszewski.... 227-229
Hydrogen dispatch, Wroblewski's.218
Hydrogen, jet process of liquefying 266-271
Hydrogen, liquefaction of, Dewar's................. . . . 280-285
Hydrogen, liquefaction of, Pictet's 164-165
Hydrogen, liquefaction of, Wroblewski's.... 218-219
Hydrogen, Wroblewski and Olszewski's attempt to liquefy.213-214
Hydrogen, Wroblewski on critical pressure of............ 266
Ice, liquid air boiled on 335
India rubber affected by intense cold 329
India rubber band experiment......32
Isothermal expansion and contraction 68-69
Jamin... 21
Joule...... .. .. 60, 61
Joule and Thomson's experiment... 61-62
Joule's experiment 60-61
Joule-Thomson effect..269-270, 297-306
Joule-Thomson effect, negative .. 301
Kinetic energy. 31
Kirchoff..........................204
Laboratory liquid air apparatus, Linde's.....................313-316
Latent heat...................... 72-76
La Tour, Cagniard de.......... 202
La Tour's law... ....20-21, 128-129
Lavoisier....................... 94
Law, La Tour's........ 20
Layer, Crookes 80-82
Leyden University, Cailletet's pump in......... 193
Leyden University, Pictet's cycles in.............. 157-158
Liebig's account of accident with Thilorier's apparatus...... 138-139

Linde.....226, 238, 265, 300, 301, 307-319, 322
Linde's liquefaction process and apparatus.... 307-319
Liquefaction in tubes, Davy's suggestion for 126-127
Liquefaction of gases, Faraday's first work on.... .... 106, 110
Liquefaction of hydrogen, Pictet's experiment in 164-165
Liquefaction process and apparatus, Linde's 307-319
Liquefied gas receiver, Cailletet's.... 195-196
Liquid air accelerating combustion............. 332-333
Liquid air apparatus, Linde's 309-312
Liquid air as source of oxygen.316-318
Liquid air as source of power......356
Liquid air defined........ 9
Liquid air dropped into water.327-328
Liquid air, experiments with..325-337
Liquid air explosive............348-353
Liquid air, filtering......... ...326-327
Liquid air, gasification of..........332
Liquid air giving two liquids..... 220
Liquid air in Dewar bulb..........327
Liquid air in flask 325
Liquid air, medical uses of 342-343
Liquide Pictet......... ......24, 169-171
Liquid floating on a gas............21
Liquid fluorine, data of.... 278-280
Liquid helium, Dewar's production of 281
Liquid hydrogen........ 280-285
Liquid hydrogen, data of......280-283
Liquid hydrogen, Olszewski's determination of constant of.227-229
Liquids and solids, solutions of, in gases 23-24
Liquids, molecular motion in.......18
Liquid state....:....................12
Liveing, G. D.........................232
Living force..........................29
Low temperatures, applications of..................... ..338-356
Machinery, Dewar's, Royal Institution................ 239
Magnetism of oxygen...... 337
Manometer, Faraday's.........124-125
Manometers, Cailletet's........187-189
Marsh gas, liquid, as refrigerant..215
Mass...... 26
Matter, critical state of..........19-20
Matter, three forms or states of. ...11
Maxwell, J. Clerk.... 150, 204, 225, 289
Maxwell, J. Clerk, on low grade heat energy..... 289
Medical uses of liquid air......342-343
Meniscus defined ... .21
Mercury frozen by liquid air...331-332
Mercury vapor, experiment in freezing.... 253-254
Metals affected by intense cold.328-329
Metals, effect of intense cold on elasticity of 255-256, 260
Metals, effect of intense cold on elongation of...... 259-260
Metals, effect of intense cold on strength of.. 256-259
Metals, Tresca's flow of........255 256
Mixture, Thilorier's 113
Moissan and Dewar's liquefaction of fluorine 276-280
Moissan, Prof 232, 277
Molecular attraction.............11-12
Molecular death 18
Molecular motion of gases.......17-18
Molecular motion of solids..........11
Mond, Dr. Ludwig 115
Monge and Clouet............ .110
Natterer, J..19, 42, 141-147, 169, 194, 211, 213, 216
Natterer's apparatus and experiment....... 141-147
Natterer's freezing mixture 145
Natterer's thermometer........145, 211
Natterer's tube 23, 213, 216
Negative Joule-Thomson effect.....301
Nitrogen, anomalies of 88
Nitrogen, Cailletet's liquefaction of........... 184
Nitrogen dispatch, Wroblewski and Olszewski's 213
Nitrogen, solidification of, by Wroblewski and Olszewski....214
Nitrous oxide, Faraday's liquefaction of......... 111
Nitrous oxide, Natterer's liquefaction of 145
Nitrous oxide, suggested by Faraday as cooling agent............114

Northmore, Thomas..106, 110, 117-118, 121, 122, 143
Northmore, Thomas, liquefactions by ... .106, 117-118
Onnes, H. Kamerlingh......49, 270, 301
Olszewski, K..42, 145, 151, 157, 165, 168, 169, 185, 203-229, 266, 267, 301
Olszewski's and Pictet's apparatus, defect in..................223
Olszewski's determination of constants of liquid hydrogen..227-229
Olszewski's liquefaction apparatus of 1890.............. .221-226
Olszewski's liquefaction of hydrogen, approximate..............221
Olszewski's static oxygen..-.. 221-226
Oxygen, Cailletet's liquefaction of.................... 183-185
Oxygen, critical pressure of, Wroblewski and Olszewski's determination....... 216-217
Oxygen, critical temperature of, Wroblewski and Olszewski's determination 217-218
Oxygen dispatch, Wroblewski and Olszewski's.......... 211-212
Oxygen, Linde's method for producing..... 317-318
Oxygen, magnetism of....... 337
Oxygen, specific gravity determination of, Wroblewski and Olszewski's....................214
Paris, Dr. John Ayrton, and Faraday 107-109
Perkins' alleged liquefaction of air............ 116-117
Permanent gases, the six so-called.150
Pictet, Raoul....22, 24, 133, 135, 150, 151, 152-171, 185, 192, 200, 205, 220, 223, 225, 233, 264, 289
Pictet, honors received by..... ...156
Pictet's cycles used by Dewar......233
Pictet's cycles praised by Wroblewski.................. 205-206
Pictet's determination of temperature........... 167
Pictet's experiment in cold well, 340-341
Pictet's frigotherapy..........338-342
Pictet's Intellectual and Moral Philosophy.....................157
Pictet's life and character......153-155
Pictet's liquefaction of oxygen dispatch.........................161
Pictet's original liquefaction apparatus... 157-163
Pictet's liquid....... ...... 24, 169-171
Pictet's work, importance of.......168
Pleischl's lecture on Natterer's apparatus 144-146
Pneumatic Institution..............96
Potential energy....... 30
Power expended in Linde's apparatus........ ....316, 318
Power, liquid air as reservoir of....356
Pressure affecting state of matter....... 16-17
Pressure, critical...................19
Pressures, enormous, in Natterer's experiments............ ...146
Pump, Cailletet's mercury.....191-195
Pump, Faraday's 113
Pump, Pictet's....................166
Purification of chemicals by cold......... ....344 347
Purity, critical state test of....343-344
Ramsay.... ....21, 22
Reaumur thermometer scale..38, 44-45
Regnault's mercury pump.........193
Release, Cailletet's 182
Ribard's local application of intense cold 342
Ribeau, George, Faraday's employer.........................102
Regeneration of cold........... .. 299
Royal Institution of England.10, 92-99
Rumford, Count..........92-95, 118-119
Rumford's, Count, experiment in liquefaction of gases...... .118-119
Second.. 25
Self-intensive refrigeration........300
Siemens, William. 299, 300, 301
Silvered gas bulbs 247
Skating rinks, Pictet's 154-155
Solid state of matter................11
Solid carbon dioxide and Crookes layer.........................84
Solids and liquids, solutions of, in gases........ 23-24
Solids, flow of. 13
Solids, vaporization of 15-16
Solution, gaseous, utilized..........24

Solution of solids and liquids in gases.... 23-24
Solvay 265
Specific heat at constant volume and at constant pressure......64 65
Specific heat, atomic................66
Specific heat of gases............64-65
Spheroidal state.......... .. 78-84, 243
State of matter affected by pressure............ 16-17
State of matter, intermediate.13-14, 20
State of matter, volume affected by 18-21
Steel burned in liquid air..... 334
Strength of metals affected by cold............ 256-259
Stromeyer..........................122
Sulphur dioxide, liquefaction of...110
Sulphureted hydrogen, liquefaction of.................... 110-111
Surface tension 78-79
Thermodynamics, second law of.70-71
Thermometer, calorimetric 58
Thermometer, electric resistance........ 54-57
Thermometer, Fahrenheit's.........39
Thermometer, gas or air.........44-51
Thermometer, Natterer's...... 211
Thermometer scales...... ..... 37, 44
Thermometers, substances for filling.........................37, 42
Thermometer, thermo-electric...51-54
Thermometric methods, Cailletet's trials of................201-202
Thilorier ... 112, 113, 137-141, 198, 269
Thilorier's apparatus exhibited by Faraday.... 112
Thilorier's apparatus, fatal accident with 138-139, 143, 145
Thilorier's cold-blast blowpipe, 141, 198
Thilorier's experiments........137-141
Thilorier's freezing mixture..... 141
Thilorier's solid carbon dioxide....137
Thomson, Sir William...........61, 62
Thomson, Elihu 219
Thompson, Benjamin...............93
Torricellian vacuum.. ....249, 252-253
Tour, Cagniard de la..... 22
Transition phenomena..........22-23
Tresca's flow of metals 255-256
Tripler and Pictet ... .296
Tripler, Chas. E..226, 235, 255, 266, 285, 286-296, 287, 289
Tripler on low grade heat energy, 288-289
Tripler's apparatus and process 290-295
Tripler's buckets for liquid air 289, 296
Tripler's life.......... 287-289
Tubes, Faraday's bent 123-128
Vacuum, a heat insulator......244-246
Vacuum and air space bulbs, efficiency compared........ 247
Vacuum, Blenkroode's experiment illustrating utility of.....245
Vacuum bulbs or vessels 244-254
Vacuum produced by liquid hydrogen......................283-284
Vapor 63-64
Vaporization of solids.... 15-16
Villard 23, 24
Volumes, relations of, in change of state 18-21
Water and air contrasted 85-86
Water, three states of....... 11-13
Water vapor.... 64
Well, frigorific.... 339
Whisky frozen by liquid air....330-331
Witowski........... .... 55, 57
Work 25, 31-32
Wroblewski and Olszewski's apparatus 206-211
Wroblewski and Olszewski's carbon monoxide dispatch........213
Wroblewski and Olszewski's nitrogen dispatch 213
Wroblewski and Olszewski's oxygen dispatch................211-212
Wroblewski and Olszewski's oxygen liquefaction...........211 212
Wroblewski, Sigmund von. ...42, 145, 151, 157, 165, 168, 203-229, 266, 301
Wroblewski's life.............203-205
Wroblewski's liquefaction of hydrogen.... 218-219
Wroblewski on liquefaction of air.220
Zambiasi 22
Zero, absolute..... ....40 41
Zero of thermometer scales......37-38

Made in the USA
Monee, IL
04 November 2026

67391581R00223